EQUATION NUMBER	EQUATION NAME	EQUATION
(10.2)	Regression coefficient	$b = \dfrac{N(\Sigma XY) - (\Sigma X)(\Sigma Y)}{N\Sigma X^2 - (\Sigma X)^2}$
(10.3)	Intercept	$a = \overline{Y} - b\overline{X}$
BOX 10.2 STEP 17	Regression	$r = \dfrac{N\Sigma XY - (\Sigma X)(\Sigma Y)}{\sqrt{[N\Sigma X^2 - (\Sigma X)^2][N\Sigma Y^2 - (\Sigma Y)^2]}}$
(10.10)	t-value	$t = r\sqrt{\dfrac{N - 2}{1 - r^2}}$
(11.3)	Reliability of a test scored 0 or 1 (KR_{20})	$KR_{20} = \dfrac{N}{N - 1}\left(\dfrac{S^2 - \Sigma pg}{S^2}\right)$
(11.4)	Coefficient alpha	$\alpha = \left(\dfrac{N}{N - 1}\right)\left(\dfrac{S^2 - \Sigma S_i^2}{S^2}\right)$
(11.5)	Prophecy formula	$N = \dfrac{r_d(1 - r_o)}{r_o(1 - r_d)}$
(11.6)	Maximum correlation of reliabilities	$r_{12}\,max = \sqrt{r_{11}r_{22}}$
(12.1)	Chi-square	$\chi^2 = \sum\dfrac{(O - E)^2}{E}$
(12.2)	Degrees of freedom	$df = $ number of categories $- 1$
(12.3)	Chi-square in a 2 × 2 table	$\chi^2 = \dfrac{N(AD - BC)^2}{(A + B)(C + D)(A + C)(B + D)}$
(12.4)	Mann–Whitney U-test	$U_1 = N_1N_2 + \dfrac{N_1(N_1 + 1)}{2} - R_1$

Basic Statistics for the Behavioral Sciences

Basic Statistics for the Behavioral Sciences

ROBERT M. KAPLAN
San Diego State University and
University of California, San Diego

ALLYN AND BACON, INC.
Boston • London • Sydney • Toronto

Managing Editor   •   Bill Barke
Developmental Editor   •   Allen Workman
Editorial / design / production service   •   Quadrata, Inc.
Copy Editor   •   Mary Ellen Gray
Text Designer   •   Geri Davis, Quadrata, Inc.
Photo Researcher   •   Yvonne Gerin and Quadrata, Inc.
Cover Coordinator   •   Linda Dickinson
Cover Designer   •   Richard Hannus
Production Editor   •   Peter Petraitis

Library of Congress Cataloging-in-Publication Data

Kaplan, Robert M.
  Basic statistics for the behavioral sciences.

  1. Psychometrics.   2. Medical statistics.   I. Title.
BF39.K27   1987   150'.72   86-1188
ISBN 0-205-08693-4

PHOTO CREDITS

2, Sybil Shelton, © Peter Arnold, Inc.   **10,** (a) Texas Instruments; (b) and (c) IBM   **14, 30,** Georg Gerster, Photo Researchers, Inc.   **40,** Robert Isaacs, Photo Researchers, Inc.   **53,** JGH, *Wakefield Daily Item*   **62,** Georg Gerster, Photo Researchers, Inc.   **68,** © Jim Anderson, Woodfin Camp and Associates   **80,** Sybil Shelton, © Peter Arnold, Inc.   **98,** Erika Stone, © Peter Arnold, Inc.   **104,** John Blaustein, Woodfin Camp and Associates   **134,** Gilles Peress, © Magnum Photos, Inc.   **157,** © Henri Cartier Bresson, Magnum Photos, Inc.   **168,** © Craig Aurness, 1981, Woodfin Camp and Associates   **190, 207, 212, 242,** Georg Gerster, Photo Researchers, Inc.   **249,** Paul Iglesias, *Boston Globe*   **264,** Georg Gerster, Photo Researchers, Inc.   **274,** Inger McCabe, Photo Researchers, Inc.   **298,** Ellis Herwig, Taurus Photos

Contents

Inferential Statistics   Part II

5 Introduction to Probability   81

6 Introduction to Inferential Statistics   105

7 Methods for Comparing Two Sets of Observations   135

8 The One-Way Analysis of Variance   169

9 The Two-Way Analysis of Variance   191

Other Statistical Techniques   Part III

10 Correlation and Regression   213

11 Reliability and Validity   243

Preface

This book was written to provide a general introduction to statistics for the behavioral and health sciences. Beyond providing instruction in techniques, I introduce the concepts of statistical thinking. Because very few things in our lives are known for certain, statistical methods help students to think about the world in a probabilistic way and to understand consistencies and inconsistencies. Through statistical analysis they can discover how to attach a degree of certainty to statements about events and observations.

Another major aim of the book is to present the fundamentals of statistics to students who have virtually no mathematics background. The objective is to develop students' confidence and understanding in applying basic statistical procedures to concrete research situations in the behavioral and health sciences. For each key procedure, this text walks students through a step-by-step presentation of basic procedures, worked out in extremely specific detail. (These procedures are further reinforced for students through the Study Guide and its accompanying software programs.) At the same time the overall framework of research-design priorities is always kept in view to help students organize the underlying reasoning needed for statistical testing of hypotheses or exploration of data.

Although the book was written primarily for students in the behavioral sciences, most of the examples come from fields in which I have found students to be interested. For example, data from professional sports, criminal behavior, family interactions, and entertainment are used in many of the examples. In my teaching of introductory statistics, I have found that college men and women generally appreciate real numbers generated from public spectacles or everyday events.

To assist students in gaining confidence in statistical reasoning and basic procedures, a distinctive Study Guide is offered. It is accompanied by a software disk of programs that students can use for ease of computation on problems in the text and Study Guide. One of the programs also helps evaluate the research-design conditions that are appropriate for the major statistical procedures.

Like most authors I expected the completion of the manuscript to be a brief project. However, things became more complicated. Work on the project was woven between 11 research grants and a maze of other administrative commitments. What began with a burst of enthusiasm during one summer vacation emerged as this book three years and two children later.

Many friends, colleagues, and students contributed to this effort. I am particularly indebted to the staff at the Center for Behavioral Medicine at San Diego State University. Connie Toevs made major contributions during the early phases of the project. She typed portions of the manuscript, wrote the computer programs for the workbook, and provided many other invaluable forms of assistance. Therese Cauchon made a variety of other contributions including the assemblage of permissions. Char Hook also did her share of the typing. In the later phases, Sandra Silva and Joyce Garman did a superb job of keeping many aspects of the project on track. Several San Diego State students, including Mary Bulcao, David Dickason, and Tom Meyer, read the manuscript and provided feedback.

Authors often complain about their publishers. Yet I can say with some sincerity that the relationship with Allyn and Bacon has been both amicable and enjoyable. Bill Barke was entertaining, supportive, and encouraging throughout the project. Allen Workman was consistently thorough and conscientious in helping to shape the manuscript. I am very thankful to professors Charles Hinderliter of the University of Pittsburgh at Johnson, Bill McDaniel of Georgia College, Robert Levy of Indiana State University, and especially Mike Brown of Pacific Lutheran University for their many helpful suggestions in the development of the project.

Geri Davis and Martha Morong of Quadrata, Inc., provided the editing and production services for the manuscript. They were a delight to work with and all aesthetic accomplishments of the textbook should be attributed uniquely to them. Finally, for the last several years the many versions of this manuscript shared a small house with my wife, Dr. Catherine Atkins, and two sons, Cameron and Seth Kaplan. They join me in the celebration of its completion.

R.M.K.

Basic Statistics for the Behavioral Sciences

1

Introduction

Using Numbers

Mathematics is a perfectly unnatural field of study. Our species created numbers, and we may be the only living beings who consciously use and manipulate them. Unlike food and water (which are used by other animal species) we have no physiological need for numbers. Yet we have become increasingly dependent on numbers, arithmetic, and mathematics to understand the world we live in. Since numbers were created by humans, they obey well-defined sets of rules.

Quantification is the description of some quality or characteristic in terms of quantities or numbers. Quantitative studies pertain to measurement or use of measurement.

Today we could hardly survive without numbers. We count, use money, measure things, and so on. You might stop and think about how many times you use numbers each day. Numbers and quantification provide us with a very special language that allows us to express ourselves precisely. Scientific investigation requires precision, and this is gained through use of quantitative methods. These methods are human-made and precise, and follow formal logic. The precision of all other sciences is linked to mathematics and the ability to describe precise logical relationships.

We constantly process information about the world. In fact, there is so much information available that we cannot organize and interpret all of it. To make more sense of the world we must measure phenomena and be able to summarize what we have measured. Methods that are used to organize, summarize, and describe observations are called **descriptive statistics.** These might include the batting averages for Boston Red Sox players, the average heights and weights for second-graders in the Chicago school system, and the association between watching television violence and behaving aggressively. In addition to descriptive statistics we also use **inferential statistics** for drawing general conclusions about probabilities on the basis of a sample. There are many uses for inferential statistics. In some cases we might use statistical inference to make statements about the eating habits of American citizens on the basis of the study of a small fraction of the American population. In other instances we might use inferential statistics to decide whether to attribute differences between two groups to chance or nonchance factors. For instance, suppose one group is given a treatment and another group is not treated. If small differences between the groups are observed, we need statistical methods to help us decide whether to attribute the differences to chance or to the treatment. In this book we will provide the basis for both descriptive and inferential statistical methods.

Measurement and Methods for Describing the World

Measurement requires the application of rules for assigning numbers to objects. The rules are the specific procedures for transforming qualities of attributes into numbers (Nunnally, 1978). For example, to rate the quality of wines, the wine taster must be given a specific set of rules. The wine might be rated on a 20-point rating scale in which 1 means extremely bad and 20 means extremely good. The basic feature of the system is the scale of measurement. For example, to measure the height of your classmates you might use a scale of inches; to measure their weight, you might use a scale of pounds.

In the behavioral and health sciences we do not have widely accepted scales of measurement. Nevertheless we do have numerous systems by which we assign numbers. Properties and characteristics of measurement are usually studied in courses on psychometrics and testing (see Kaplan and Saccuzzo, 1982, for an overview).

Games, Gods, and Gambling— The History of Statistics and Probability Theory

Historians have been unable to determine precisely when humankind began using number systems. There is speculation, however, that the earliest number systems were associated with the number of fingers and thumbs on the two hands. Base-ten number systems (representing our ten fingers) are universal systems of counting for all civilized people and nearly all primitive tribes. The ancient Persians and Greeks had words for five that meant "hand." Some speculate that the Roman numeral V is for the V between the thumb and fingers of one hand. Early counting systems may have used the several joints of the fingers for notational systems. Placing the hand at certain positions on the human body expressed multiples of five and ten. By the eighth century an English historian and teacher called the "venerable Bede" was able to use the system to count to one million (David, 1962). By the ninth or tenth century people in India were writing down numbers as we do today. Yet progress in the development of arithmetic was slow for cultures such as the Greeks and Romans because their notation was very cumbersome. Arabic notation made progress in arithmetic and mathematics easier.

As civilizations progressed they became more fascinated with games and gambling. There is evidence from the Babylonians, Egyptians, Greeks, and Romans of the pre-Christian era that games of chance were prevalent. Greek vase paintings have shown boys playing a marblelike game of chance. There is also evidence from the first dynasty in Egypt (3500 B.C.) from tomb paintings depicting board games. In sum, recorded history reveals a long trail of interest in games of chance and gambling. To become better gamblers, however, they had to learn more about the rules of probability.

Rolling dice is one form of gambling that has been around for many centuries, perhaps as early as 3000 B.C. The earliest dice date to the beginning of the third millennium. There is evidence that dice games were popular among many cultures of the Mediterranean region. Some have conjectured that the dots on die representing the numbers one through six were used instead of other number systems because they were easier to carve into pottery. Over the centuries, many distinguished philosophers and mathematicians have studied outcomes of dice games. The probabilities of events in dice games were considered by Sir Isaac Newton, James Bernoulli, Galileo, Leonardo da Vinci and the French philosopher–mathematician Blaise Pascal. In fact, in the middle of the seventeenth century Pascal worked out rules of binomial power, which are essential to probability theory, in a series of letters to Frenchman Pierre de Fermat. Pascal's method is known as the Pascal triangle.

The development of games and gambling forced scholars and gamblers alike to understand the probability that events would occur by chance. A sophisticated gambler by the eighteenth century might understand the probability that his risk would pay off. In this book, you will learn about statistical methods for estimating the likelihood that observed events are more common than chance. A flip of a coin will come up heads half of the time. Similarly, a student given a true-false test will get half of the problems correct by chance. Yet would we call it chance if a student got 65% of the items correct? Statistical methods help us decide whether performance is significantly better than chance.

The roots of modern statistics are not only in games and gambling. Statistical methods were also developed to tabulate numerical data. Although interest in probability theory was keen among Italian and French philosophers, it apparently did not interest English mathematicians. Florence David, the British probability theory historian, noted that while probability theory was developing on the European continent in the sixteenth and seventeenth centuries "the English, as usual, were preoccupied with concrete facts" (David, 1962, p. 98).

In England, statistical methods began with enumeration of various matters for the church. Thomas Cromwell (1485–1540) first introduced the registration of all weddings, christenings, and burials. The first health statistics were introduced in 1594 as "Bills of Mortality." They were listings of a number of plague deaths in comparison to the deaths from other sicknesses. However, the bills were discontinued in 1595 after a plague epidemic wiped out a sizeable portion of the British population. They were reintroduced in 1603, and by 1625 weekly bills of mortality were available. A mortality summary was published annually on the Thursday before Christmas; the summary for 1665 appears in Fig. 1.1. As the bill shows, plague was by far the most common ailment of the era (68,596 cases).

During this period, the purpose of keeping vital statistics was to understand the intentions of God. Florence Nightingale wrote

> The true foundation of theology is to ascertain the character of God. It is by the aid of statistics that law in the social sphere can be ascertained and codified, and certain aspects of the character of God thereby revealed. The study of statistics is thus a religious service. (In David, 1962, p. 103)

In fact many of the early statistical analyses reported the religious view. Perhaps the first vital statistician was John Graunt who published a book entitled *Natural and Political Observations on the Bills of Mortality* (1662). The book marks the first attempt to find determinants of the cause of death. Table 1.1 describes some of Graunt's analysis. As the table shows, the percentage of all deaths attributable to the plague varies across years. The worst year was 1603. Also presented in the table is the percentage of people christened in each of the years. You may also notice that 1603 was not only the year that plague deaths rose to their highest level but was also the year that the lowest percentage of christenings occurred. Although some believed the plague was God's punishment for the lack of christenings, the cause and effect were not clearly established. During the plagues, many people took their children away from the cities to avoid the infectious disease. Thus, the children could not come to the church to be christened.

Modern statistical methods began to emerge toward the end of the nineteenth century. Charles Darwin was a keen observer of behavior and used intuitive notions of probability and statistics in his highly influential *Origin of Species,* published in 1859. Darwin's cousin, Sir Francis Galton, soon began applying Darwin's ideas to the study of individual differences among humans. Galton was the first to describe what later was known as regression (see Chap-

Figure 1.1

Example of a yearly bill of mortality.

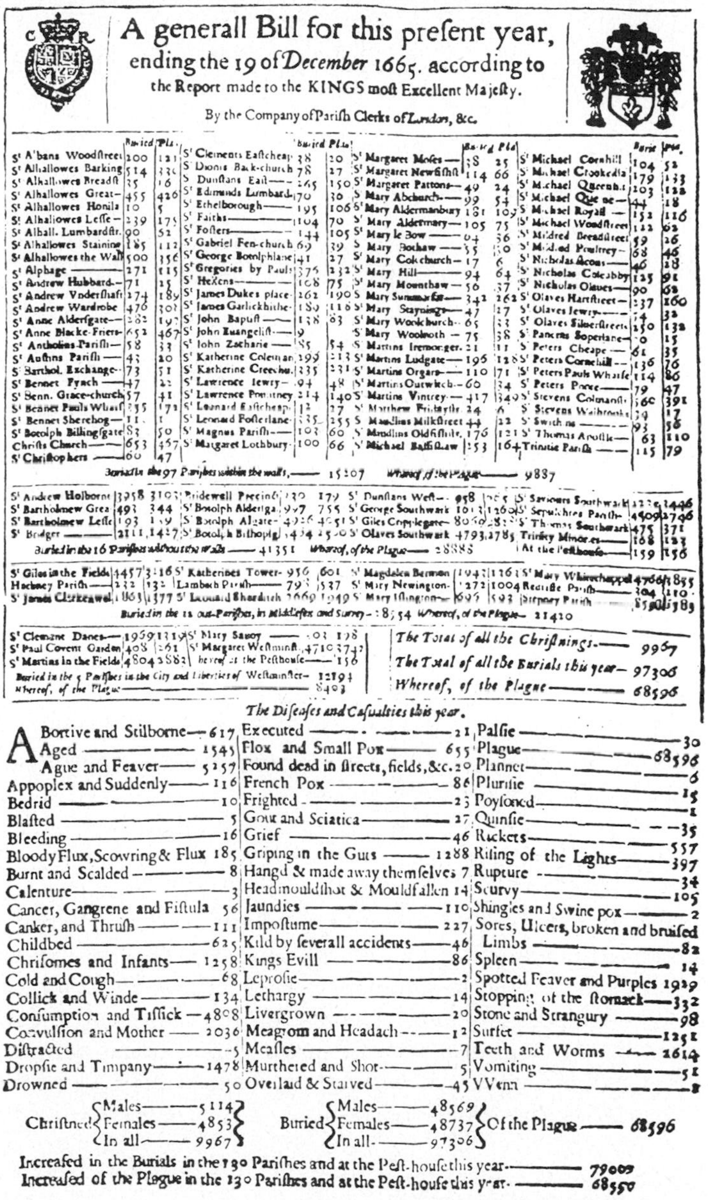

A generall Bill for this present year, ending the 19 of *December* 1665. according to the Report made to the KINGS most Excellent Majesty.

By the Company of Parish Clerks of *London*, &c.

	Buried	Plague		Buried	Plague		Buried	Plague		Buried	Plague
St Albans Woodstreet	200	121	St Clements Eastcheap	38	20	St Margaret Moses	38	25	St Michael Cornhill	104	52
St Alhallowes Barking	514	330	St Dionis Back-church	78	27	St Margaret Newfishst	114	66	St Michael Crookedla	179	133
St Alhallowes Breadst	35	16	St Dunstans East	265	150	St Margaret Pattons	49	24	St Michael Queenhi	203	122
St Alhallowes Great	455	426	St Edmunds Lumbard.	70	36	St Mary Abchurch	99	54	St Michael Querne	44	18
St Alhallowes Honila	10	5	St Ethelborough	195	106	St Mary Aldermanbury	181	109	St Michael Royall	152	116
St Alhallowes Lesse	239	175	St Faiths	104	70	St Mary Aldermary	105	75	St Michael Woodstreet	122	62
St Alhall. Lumbardstr.	90	52	St Fosters	144	105	St Mary le Bow	64	36	St Mildred Breadstreet	59	26
St Alhallowes Staining	185	112	St Gabriel Fen-church	69	39	St Mary Bothaw	55	30	St Mildred Poultrey	68	46
St Alhallowes the Wall	500	356	St George Botolphlane	41	27	St Mary Colechurch	17	6	St Nicholas Acons	46	28
St Alphage	271	115	St Gregories by Pauls	376	232	St Mary Hill	94	64	St Nicholas Coleabby	125	91
St Andrew Hubbard	71	25	St Helens	108	75	St Mary Mounthaw	56	37	St Nicholas Olaves	90	62
St Andrew Vndershaft	274	189	St James Dukes place	262	190	St Mary Summerset	342	262	St Olaves Hartstreet	237	160
St Andrew Wardrobe	476	308	St James Garlickhithe	189	118	St Mary Staynings	47	27	St Olaves Iewry	54	32
St Anne Aldersgate	282	197	St John Baptist	138	83	St Mary Woolchurch	65	33	St Olaves Silverstreet	250	132
St Anne Blacke-Friers	652	467	St John Euangelist	9		St Mary Woolnoth	75	38	St Pancras Soperlane	30	15
St Antholins Parish	58	33	St John Zacharie	85	54	St Martins Iremonger.	21	11	St Peters Cheape	61	35
St Austins Parish	43	20	St Katherine Coleman	299	213	St Martins Ludgate	196	128	St Peters Cornehill	136	76
St Barthol. Exchange	73	51	St Katherine Creechu.	335	231	St Martins Orgars	110	71	St Peters Pauls Wharfe	114	86
St Bennet Fynch	47	22	St Lawrence Iewry	94	48	St Martins Outwich	60	34	St Peters Poore	79	47
St Benn. Grace-church	57	41	St Lawrence Pountney	214	140	St Martins Vintrey	417	349	St Stevens Colmanst.	560	391
St Bennet Pauls Wharf	355	172	St Leonard Eastcheap	42	27	St Matthew Fridaystr.	24	6	St Stevens Walbrooke	34	17
St Bennet Sherehog	11	1	St Leonard Fosterlane	335	255	St Maudlins Milkstreet	44	22	St Swithins	93	56
St Botolph Billingsgate	83	50	St Magnus Parish	103	60	St Maudlins Oldfishstr.	176	121	St Thomas Apostle	163	110
Christs Church	653	467	St Margaret Lothbury	100	66	St Michael Bassishaw	253	164	Trinitie Parish	115	79
St Christophers	60	47									

Buried in the 97 Parishes within the walls, — 15207 *Whereof, of the Plague* — 9887

St Andrew Holborne	3958	3103	Bridewell Precinct	230	179	St Dunstans West	958	665	St Saviours Southwark	4235	3446
St Bartholmew Great	493	344	St Botolph Aldersga.	997	755	St George Southwark	1613	1260	St Sepulchres Parish	4509	2746
St Bartholmew Lesse	193	139	St Botolph Algate	4926	4051	St Giles Cripplegate	8069	4838	St Thomas Southwark	475	371
St Bridget	2111	1427	St Botolph Bishopsg.	3464	2500	St Olaves Southwark	4793	2785	Trinity Minories	168	123
									At the Pesthouse	159	156

Buried in the 16 Parishes without the walls — 41351 *Whereof, of the Plague* — 28888

St Giles in the Fields	4457	3216	St Katherines Tower	956	601	St Magdalen Bermon	1943	1363	St Mary Whitechappel	4766	3855
Hackney Parish	232	132	Lambeth Parish	798	537	St Mary Newington	1272	1004	Redriffe Parish	304	210
St James Clerkenwel	1863	1377	St Leonard Shorditch	2669	1949	St Mary Islington	696	593	Stepney Parish	8598	6583

Buried in the 12 out-Parishes, in Middlesex and Surrey — 28554 *Whereof, of the Plague* — 21420

St Clement Danes	1969	1319	St Mary Savoy	303	198
St Paul Covent Garden	408	261	St Margaret Westminst.	4710	3742
St Martins in the Fields	4804	2883	hereof at the Pesthouse		156

Buried in the 5 Parishes in the City and Liberties of Westminster — 12194
Whereof, of the Plague — 8403

The Total of all the Christnings — 9967
The Total of all the Burials this year — 97306
Whereof, of the Plague — 68596

The Diseases and Casualties this year.

Abortive and Stilborne	617	Executed	21	Palsie	30
Aged	1545	Flox and Small Pox	655	Plague	68596
Ague and Feaver	5257	Found dead in streets, fields, &c.	20	Plannet	6
Appoplex and Suddenly	116	French Pox	86	Plurisie	15
Bedrid	10	Frighted	23	Poysoned	1
Blasted	5	Gout and Sciatica	27	Quinsie	35
Bleeding	16	Grief	46	Rickets	557
Bloody Flux, Scowring & Flux	185	Griping in the Guts	1288	Rising of the Lights	397
Burnt and Scalded	8	Hangd & made away themselves	7	Rupture	34
Calenture	3	Headmouldshot & Mouldfallen	14	Scurvy	105
Cancer, Gangrene and Fistula	56	Jaundies	110	Shingles and Swine pox	2
Canker, and Thrush	111	Impostume	227	Sores, Ulcers, broken and bruised Limbs	82
Childbed	625	Kild by severall accidents	46	Spleen	14
Chrisomes and Infants	1258	Kings Evill	86	Spotted Feaver and Purples	1929
Cold and Cough	68	Leprosie	2	Stopping of the stomack	332
Collick and Winde	134	Lethargy	14	Stone and Strangury	98
Consumption and Tissick	4808	Livergrown	20	Surfet	1251
Convulsion and Mother	2036	Meagrom and Headach	12	Teeth and Worms	2614
Distracted	5	Measles	7	Vomiting	51
Dropsie and Timpany	1478	Murthered and Shot	5	VVenn	8
Drowned	50	Overlaid & Starved	45		

Christned: Males — 5114; Females — 4853; In all — 9967

Buried: Males — 48569; Females — 48737; In all — 97306. Of the Plague — 68596

Increased in the Burials in the 130 Parishes and at the Pest-house this year — 79009

Increased of the Plague in the 130 Parishes and at the Pest-house this year — 68590

Source: F. David, *Games, Gods and Gambling*. New York: Hafner, 1962.

Table 1.1					
Example of Graunt's analysis in an attempt to find causes of death					
PERIOD	ALL BURIALS	PLAGUE VICTIMS	% PLAGUE	CHRISTENED	(CHRISTENED/ PLAGUE) %
1592	26,490	11,503	43	4,277	37
1593	17,844	10,662	60	4,021	38
1603	42,042	36,269	86	4,784	13
1625	54,265	35,417	65	6,983	20
1636	23,359	10,400	45	9,522	92

Source: F. David, *Games, Gods, and Gambling.* New York: Hafner, 1962.

ter 10). The product moment correlation (see Chapter 10) was developed by Karl Pearson and published in 1896. Pearson introduced methods for descriptive statistics that are still used today. The next major advance in statistics was the publication of Sir Ronald Fisher's book, *Statistical Methods for Research Workers* in 1936. In this book, Fisher derived many statistical methods that form the basis for many of our current procedures.

An Overview of the Book

This book will present some of the basic methods for understanding and performing research. The book is divided into three parts: descriptive statistics, inferential statistics, and statistical techniques.

■ Descriptive Statistics

These are methods that concisely describe a collection of quantitative information. In Chapter 2 you will learn about graphs and distributions. Different measurement scales will be used to summarize quantitative information in graphs. In Chapter 3 you will learn to describe characteristics of the distributions using summary statistics. In Chapter 4 you will learn to find points within distributions of scores. In this process you will use percentiles and specific mathematical indexes known as standard scores.

■ Inferential Statistics

These are methods used in making inferences from observations of a small group, known as a **sample,** to a larger group, known as the **population.** Typically we want to make statements about the larger group, but it is not possible to make all of the necessary observations. Instead observations are made on a relatively small group of subjects, and inferential statistics are used to estimate the characteristics of the larger group. This part of the book begins with a discussion of elementary probability theory in Chapter 5. This will help you understand some of the basic rules of probability that were first worked out by the early philosophers/gamblers. Then you will learn the basics of statistical hypothesis testing. Chapter 6 states rules that you will eventually apply in deciding whether events are chance or nonchance. This discussion is expanded in Chapter 7, which reviews a specific set of methods known as the t-test. More advanced methods for comparing groups, known as the analysis of variance, are presented in Chapters 8 and 9.

■ Other Statistical Techniques

Chapter 10 covers methods for simultaneously describing two distributions. You will first learn to draw pictures of two distributions, and then you will learn how to apply mathematical indexes to describe the relation between the two. These methods can sometimes be abstract. Therefore, an entire chapter (Chapter 11) will illustrate applications of the correlation and regression indexes that are featured in Chapter 10. Most of the methods covered in the first eleven chapters of the book deal with scaled quantitative data. In Chapter 12, you will learn about nonparametric statistics, which are used to study measures expressed in less refined scales. For example, nonparametric statistics are used in the study of proportions and ranks. As you can see, the book covers many different topics. Thus, we have included Chapter 13 as a general wrap-up. Chapter 13 attempts to put together many of the different terms, concepts, and methods you will learn in your course.

Computers and Calculators

To learn statistics efficiently you will need many tools. Three different tools for analyzing data will be suggested in this book, and exercises for using them are presented in the workbook. The first tool is a hand-held calculator. Hand-held calculators, as shown in Fig. 1.2(a), have emerged in great numbers since around 1970. These exceptional little machines have revolutionized the use of statistical methods. Many of today's calculators can perform most of the analyses described in this book. However, they are not generally efficient at storing data or manipulating data sets. For these purposes you may need a computer.

Microprocessors are perhaps the most important innovation of the 1980s. Small computers such as those shown in Fig. 1.2(b) are very efficient at making rapid calculations and are found in an increasing number of homes and colleges. Microprocessors have the advantage of storing data sets, manipulating

Figure 1.2

Calculators and computers.

(a)

(c)

(b)

data, and performing complex calculations. Most of the calculations required for behavioral research can be accomplished with a microprocessor. However, some data sets are too large to be handled easily by a small computer.

For very large data sets we often require the services of a mainframe computer (see Fig. 1.2c). These large computers are capable of very rapid calculations on enormous data sets. To accomplish these calculations large, complex computer programs are required. There are a variety of excellent programs available that accomplish all of the analyses we will discuss in this book. For example, Statistical Package for Social Science (SPSS) program (Nie, et al., 1966; Hull and Nie, 1981) is a series of routines designed specifically for statistical analysis in the social, behavioral, and health sciences. It is well documented, efficient, and available at nearly all university computer centers. Although SPSS is not covered in this book, students with a serious interest in statistics will benefit from learning to use it.

Exercises

MULTIPLE CHOICE

1. Inferential statistics
   a. are used to describe a population of scores.
   b. are used only to compare an experimental and control group.
   c. are used for drawing general conclusions of a probabilistic nature on the basis of a sample.
   d. were used in the "bills of mortality."

2. Probability theory most likely grew out of experiences with
   a. religion.
   b. gambling.
   c. horse racing.
   d. the nuclear arms race.

3. The first recorded use of statistics in England was
   a. Pascal's triangle.
   b. as bills of mortality.
   c. for experimental studies on the plague.
   d. to record records of British athletes.

Descriptive Statistics

I

Throughout your lifetime, you will continually accumulate numerical information. It is simply unavoidable. Each day, the newspaper is filled with numerical information. The first section of this book is about descriptive statistics. The purpose of descriptive statistics is to organize or to summarize numerical information.

Numerical information is often organized or summarized in the form of graphs. Methods for creating graphs are discussed in Chapter 2. Chapter 2 also describes the use of frequencies and proportions to help us understand and summarize information.

Another form of numerical description is the computation of average scores. Averages are described in Chapter 3. In addition, Chapter 3 provides methods for describing the variability among sets of scores.

The final chapter in this section gives methods of finding points within distributions of scores. These methods help us understand whether particular scores are common or unusual.

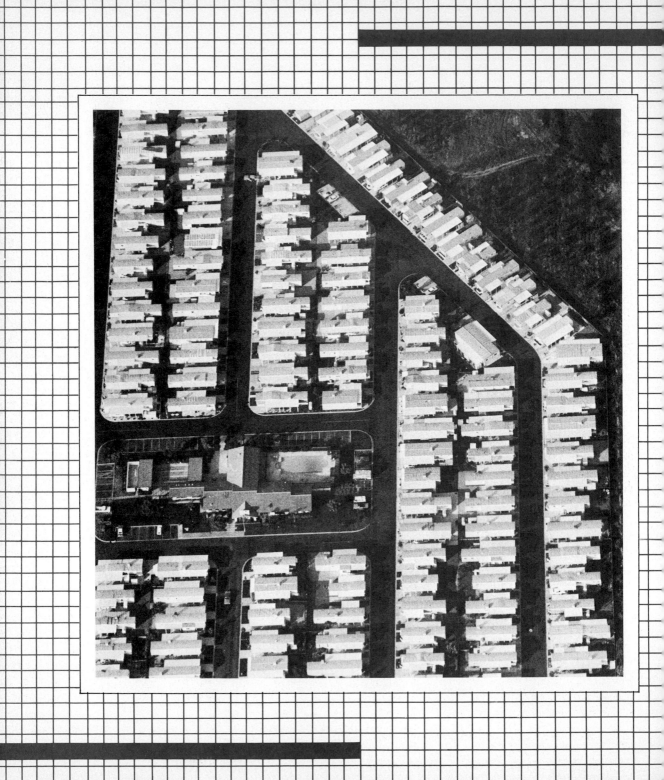

Graphs and Their Distributions

Take out your driver's license and look at all the numbers on it. It probably lists your height and weight in addition to your unique driver's license number. Although height, weight, and license number are all expressed in numerals, the numbers have quite different meanings. The unique driver's license number is for identification; it is yours and cannot be used by any other driver. It would make no sense to average driver's license numbers or to use them in arithmetic. On the other hand, height and weight immediately provide information. A police officer can look at the height and weight on your driver's license, guess your size, and make some estimate as to whether the license belongs to you. Height and weight are measured on well-defined scales. When my driver's license reports that I am 5'10" tall, we know that the top of my head is 5'10" off the ground. It also gives us some other information because many men have been measured, and we can put our measurements into context. For example, studies of American men reveal that the average man is about 5'9" tall. When my driver's license reports that I am 5'10" tall, it suggests that I am approximately the average height for an American male.

This chapter reviews different scales of measurement. As you will see in the next section scales of measurement like driver identification number represent *nominal* scales of measurement. Information such as "I am shorter than my friend Tom, and Tom in turn is shorter than Ernie" form an *ordinal* scale of measurement because it orders us according to height. It is important for you to learn about these scales of measurement, because they define the types of mathematical manipulation that are permissible.

After discussing scales of measurement we present methods for drawing pictures of data. Sometimes it is hard to make sense out of masses of numbers, but by organizing the information in the form of a graph it is often easier to make sense out of it.

Before presenting methods for drawing pictures of data (graphs), we review the characteristics of numerical scales. Knowing about these scales is not only important for drawing graphs, but it is necessary for choosing the right method of statistical analysis and making the appropriate interpretations of numerical data.

Properties of Scales

There are three important properties that make scales of measurement different from one another: magnitude, equal intervals, and absolute 0.

■ Magnitude

Magnitude is the property of "moreness." A scale has the property of magnitude if we can say that one attribute is more than, less than, or equal to another attribute (McCall, 1980). On a scale of height, for example, if we can say that Bill is taller than Mike, then the scale has the property of magnitude. An example of a scale that does not have the property of magnitude is the numbers on the uniforms of baseball players; here the numbers are only used to label the players. If the coach were to rank order the players by their batting averages, the new numbering system (batting average) would have the property of magnitude.

■ Equal Intervals

The concept of equal intervals is a little more complex. A scale has the property of equal intervals if the difference between two points is uniform along the entire scale. For example, the difference between inch 1 and inch 3 on a ruler means the same as the difference between inch 9 and inch 11. The difference is exactly 2 inches in each case.

As simple as this concept seems, a psychological test rarely has the property of equal intervals. For example, the difference between IQs of 45 and 50 does not mean the same thing as the difference between IQs of 105 and 110. Although each of these differences is 5 points, the 5 points at one level do not mean the same thing as 5 points at a higher level. When a scale has the property of equal intervals the relationship between the measured units and some outcome can be described by a straight line or a linear equation. This equation tells us that as we increase in equal units on one scale, there are equal increases in the units of another scale. The meaning of linear relationships will be covered in more detail in Chapter 10.

Table 2.1			
Scales of measurement and their properties			
	PROPERTY		
TYPE OF SCALE	Magnitude	Equal Interval	Absolute 0
Nominal	No	No	No
Ordinal	Yes	No	No
Interval	Yes	Yes	No
Ratio	Yes	Yes	Yes

Source: Kaplan and Saccuzzo, 1982.

■ Absolute 0

An absolute 0 is obtained when nothing at all exists of the property being measured. For example, if you are measuring wind velocity, but there is a reading of 0, you would conclude that there is no wind at all. For many psychological qualities it is extremely difficult, if not impossible, to define an absolute 0 point. For example, if we are measuring musical aptitude on a rating scale of 0 through 10, it is hard to say that 0 means the person has no aptitude at all. It is possible that there is some level of ability the scale does not measure or that there is no such thing as zero ability. We usually assume that if someone is alive, they have "some" ability. The best example of absolute 0 comes from chemistry. Absolute 0 on the Kelvin scale of temperature is a point at which all molecular activity stops.

Combining these three properties—magnitude, equal interval, and absolute 0—allows us to describe four different scales of measurement. These scales and their properties are defined in Table 2.1.

Types of Scales

Table 2.1 shows that a **nominal scale** does not have the property of magnitude, equal intervals, or absolute 0. Nominal scales are really not scales at all; their only purpose is to name objects. For example, the numbers on baseball players' uniforms are nominal in nature. In social science research, groups in sample surveys are commonly labeled with numbers (such as 1 = white, 2 = Black, and 3 = Mexican-American). However, when these numbers have been attached to categories, averaging the numbers together is not usually advisable. On the scale above for ethnic groups, the average score of 1.87 would have no meaning.

A scale with the property of magnitude but not the property of equal intervals or the property of absolute 0 is known as an **ordinal scale**. An ordinal scale allows us to rank individuals or objects but not to say anything about the meaning of the differences between the ranks. If you were to order the members of your class in terms of their height, you would have an ordinal scale. Note that you would do this without concern for the differences between the ranks. For example, if Connie were the tallest, Carma the second tallest, and Cathie the third tallest, you would assign them the ranks of 1, 2, and 3, respectively. You would not consider the fact that Connie is 8 inches taller than Carma but that Carma is only 2 inches taller than Cathie.

For most problems in psychology we do not have the precision to measure the exact differences between intervals, so most often we use ordinal scales of measurement. For example, IQ tests do not have the property of equal intervals or absolute 0, but do have the property of magnitude. If they had the property of equal intervals, the difference between an IQ of 70 and one of 90 would have the same meaning as the difference between an IQ of 125 and one of 145. Because it does not, the scale can only be considered ordinal. Furthermore there is no point on the scale at which there is no intelligence at all. Thus, the scale does not have the property of absolute 0. (See Kaplan and Saccuzzo, 1982, for a detailed discussion of IQ.)

When the scale has the property of magnitude and equal intervals but not the property of an absolute 0, we refer to it as an **interval scale**. The most common example of an interval scale is the measurement of temperature in degrees Fahrenheit. This temperature scale clearly has the property of magnitude since 35°F is warmer than 32°F, 65°F is warmer than 64°F, and so on. Also, the difference between 90°F and 80°F is equal to a similar difference of 10°F at any point on the scale. However, on the Fahrenheit scale temperature does not have the property of an absolute 0. If it did, the 0 point would be more meaningful. As it is, 0 on the Fahrenheit scale does not have a particular meaning. Freezing occurs at 32°F, and boiling water occurs at 212°F. Because this scale does not have an absolute 0, it is not possible to make statements in terms of ratios. A temperature of 22°F is not twice as hot as 11°F, and a temperature of 70°F is not twice as hot as one of 35°F. The Celsius scale of temperature is also an interval rather than ratio scale. Although 0 represents freezing on the Celsius scale, the 0 is still not an absolute 0. (Remember that absolute 0 is a point at which nothing of the property being measured exists.) Even on the Celsius scale of temperature there is still plenty of room on the thermometer below 0. When the temperature goes below freezing, some temperature is still being measured.

A scale that has all three properties (magnitude, equal interval, and absolute 0) is called a **ratio scale**. To continue our example with the scales of temperature, a ratio scale is one that has the properties of the Fahrenheit scale and the Celsius scale but also includes a meaningful 0 point. Physicists and chemists tell us that there is a point at which all molecular activity ceases. This

Scales of Measurement Review

The left column contains a series of questions about scales of measurement. The answers and explanations are in the right column. Use the Box for review, or use a piece of paper to cover the right side of the box and see how many of the questions you can get correct.

EXAMPLE

ANSWERS AND EXPLANATIONS

1. Teams in the intramural basketball league are numbered 1 through 10, that is, Team 1, Team 2, ... What type of measurement is this?

Nominal

2. Jane's mother claims that Jane is twice as sick today as she was yesterday. Yesterday, Jane's temperature was 2° above normal, and today her temperature is 4° above normal. Is Jane's mother correct in claiming she is twice as sick?

No, the statement, "twice as sick" implies a ratio scale of measurement. "Normal" temperature (98.6°) is an arbitrary reference point used by Jane's mother. In relation to this, 4° is twice as large as 2°. However, 102.6° is not twice as high as 100.6°. Most likely, Jane's mother is using an interval scale.

3. What type of scale is represented by the list of ten fastest women marathon runners?

Ordinal

4. Poor Bill scored 0 on the test of manual dexterity. Does this mean that Bill has no dexterity at all?

Probably not. The score of 0 on the dexterity test is most likely an arbitrary 0. If Bill has hands and uses them, he probably has some dexterity that the test is incapable of measuring.

5. What is the advantage of an interval scale over an ordinal scale?

An interval scale has the property of equal intervals. Both scales are capable of rank ordering objects. However, only the interval scale is capable of expressing the distance between measured ranks. For example, consider the height of three mountains: Mt. McKinley, Mt. Rainier, and Mt. Whitney. Mt. McKinley is 20,320 feet, Rainier is 14,910 feet, and Whitney is 14,995 feet. If we rank order them, we will find that McKinley is the tallest, Whitney the second tallest, and Rainier the third tallest. However, Mt. McKinley is 5,835 feet taller than Mt. Whitney and Whitney in turn is only 85 feet taller than Rainier. Thus, using a scale with a property of equal interval gives us much more information.

is the point of absolute 0 for temperature. The Kelvin scale of temperature includes this absolute 0 point and is thus an example of a ratio scale. There are some examples of ratio scales in the numbers we see on a regular basis. Consider the number of yards gained by running backs on professional football teams. Zero yards actually means that the player has gained no yards at all. If one player has gained 1000 yards and another has gained only 500, then we can say that the first athlete has gained twice as many yards as the second.

Another example is speed of travel. For instance, 0 mph (miles per hour) is the point at which there is no speed at all. If you are driving onto the freeway at 30 mph and you increase your speed to 60 when you get into the fast lane, it is appropriate to say you have doubled your speed.

Discrete and Continuous Variables

Another basic way to classify variables is as discrete or continuous. A **continuous variable** may take on any value within a defined range. For example, time is a continuous variable. For swimmers racing in the Olympic games, speeds are recorded down to the hundredth of a second. If finer timing equipment were available, timing could be more precise—down to the thousandth or ten-thousandth of a second. These fine divisions are possible because time is a continuous process.

Discrete variables can be either names or numbers. By definition, discrete variables inherently have gaps between successive observable values. For example, we might say that the number of fights a boxer has won is a discrete variable. The values are whole numbers of integers (1, 3, 4, 6, etc.) and not portions of numbers. Sometimes discrete variables are made up of nonoverlapping categories. Consider the discrete variable for religion that might include the categories Protestant, Catholic, Jew, and Muslim. The numbers 1, 2, 3, and 4 might be used to label these religious groups. For instance, 1 would be used for Protestant, 2 for Catholic, and so on. Fractions between discrete variables are not used.

The distinction between discrete and continuous variables is not always easy to make. For example, when we write down a value for a continuous variable we put it into a category and technically make it discrete. A swimming time of 47.28 seconds uses a discrete categorization of hundredths of a second. Despite this technical problem, the distinction between discrete and continuous variables has important practical and theoretical implications in statistics. For example, the spaces between discrete variables are not meaningful. It makes no sense to report that women have an average of 2.27 babies, yet we often do so. We return to this concept many times in this book.

Distributions and Their Graphs

A single score will mean more to us if we think about it in relation to other scores. For example, if you got a score of 31 on your statistics test, you might want to know, "Is that a good score? An average score? Does it pass?" To make sense of the information, we often place the score within a distribution of scores. In the next few chapters we review methods for obtaining actual and theoretical distributions of scores.

■ Frequency Distribution

A **frequency distribution** is a simple way of displaying and summarizing numerical information. To create a frequency distribution you need only nominal measurements. However, frequency distributions can also be made for ordinal, interval, and ratio data. A frequency distribution is defined as a presentation of data showing the frequency with which each score occurs.

Box 2.2 shows the steps in the preparation of a frequency distribution. The raw data are shown in the first portion of the box. The scores are for the first quiz in a statistics class. Students in the class are identified by their initials. In the first step each student's score on the test is recorded. For instance, the first student, SQ, received a score of 7, BT received a score of 6, and so on. In the next step the possible scores on the test are arranged in descending order. The top score possible (10) is listed at the top of the column. Now we go back to the raw data and tally the scores of each student. SQ obtained a 7 on the quiz, so a tally mark is made next to 7; BT obtained a 6, so a tally is made next to 6, and so on. The third column in the bottom half of Box 2.2 is labeled f and presents the frequency information in Arabic numerals. These are the *raw frequencies* for each score. The bottom of the frequency column says $N = 15$. In statistics we commonly use the letter N to represent the number of cases. There were 15 students who took the quiz; thus, $N = 15$ tells us that there were 15 cases in the analysis.

You have probably seen many frequency distributions such as the one in Box 2.2. Classroom teachers usually use them to summarize performance on a test. In statistical analysis a frequency distribution such as this one is most often used to summarize raw data. There are other levels at which you can summarize the data.

■ Grouped Frequency Distributions

In Box 2.2 we gave a tally for each possible score; however, there are many circumstances in which it would be impractical to tally every possible score. For

Steps in the Preparation of a Frequency Distribution for the First Statistics Quiz

STEPS

1. Record raw data.

Student (identified by initials)	Score
SQ	7
BT	6
FL	9
BH	2
SK	5
CA	6
AD	10
TC	7
JL	7
TA	5
SB	6
BG	7
AJ	8
RP	4
F I	7

2. Arrange possible scores in descending order and tally the number of students obtaining each score. There are ten items on the quiz, so the highest score possible is 10.

Score	Tally	f
10	|	1
9	|	1
8	|	1
7	|||||	5
6	|||	3
5	||	2
4	|	1
3		0
2	|	1
1		
		$N = 15$

Table 2.2

Frequency distribution for batting averages in the National League, August 1982

CLASS INTERVAL	REAL LIMITS OF INTERVAL	MIDPOINT	f
320–329	319.5–329.5	324.5	2
310–319	309.5–319.5	314.5	1
300–309	299.5–309.5	304.5	7
290–299	289.5–299.5	294.5	10
280–289	279.5–289.5	284.5	11
270–279	269.5–279.5	274.5	16
260–269	259.5–269.5	264.5	15
250–259	249.5–259.5	254.5	11
240–249	239.5–249.5	244.5	8
230–239	229.5–239.5	234.5	3
220–229	219.5–229.5	224.5	2
210–219	209.5–219.5	214.5	0
200–209	199.5–209.5	204.5	3

Note: Range = 202–321, N = 89

example, you might create a distribution that has a large number of possible scores, but in which the frequency for many of the possibilities is zero. A few other scores might have a frequency of one, and just a few have a frequency of two or more. Under these circumstances, it is useful to group scores in intervals. A distribution of scores falling within each defined interval is then created.

The **class interval** is a portion of a measurement scale containing more than one possible value. Table 2.2 is a frequency distribution of batting averages for National League players. A batting average is a proportion between .000 and 1.000. It is common to multiply batting averages by 1000 in order to obtain whole numbers. For example, a batting average of .250 is usually called 250. Batting averages greater than 333 are rare. Nevertheless, listing every obtained batting average between say, 150 and 333 would make for a very large and awkward frequency distribution. To make a neater summary, class intervals of 10 points have been created. The top interval is for averages of 320–329; the next one is 310–319; and so on. How many intervals should be used? The answer is that it is arbitrary. In this example a class interval of size 10 is chosen because it provides for 13 intervals. Distributions with 7 and 16 intervals are seen most frequently in statistics literature. The exact number of intervals is arbitrary, however, and is usually chosen to best represent the data at hand.

Upper and Lower Real Limits; Midpoints

There are several technical considerations for the proper use of grouped frequency distributions. First, the intervals have *upper real limits* and *lower real limits*. To understand this consider Table 2.2. Suppose a player has a batting average of 299.7, which falls between the class intervals.

Class intervals, or grouped frequency distributions, are typically marked by integer values, that is, they are defined by whole numbers. However, the lower real limit of an interval extends halfway to the upper limit of the next class interval. Thus the interval described as 300–309 has as its real values 299.5–309.5. Similarly, the interval with the stated range of 290–300 has real values ranging from 289.5 to 299.5. The batting average 299.7 actually falls into the 300–309 interval.

Knowing the real limits of the class interval is necessary to find the interval midpoint. The *midpoint* is the point that is exactly halfway between the lower real limit and the upper real limit. (This will become important in Chapter 3 when you learn how to calculate mean, or average, scores from grouped data.) To find the midpoint of the interval we must first determine the exact interval size. To find the size the lower real limit of the interval is subtracted from the upper real limit. For example, in the batting average interval 290–299 the lower real limit is 289.5 and the upper limit is 299.5. Thus, the size of the interval is

$$299.5 - 289.5 = 10.$$

The midpoint of the interval is halfway between the lower real limit and the upper real limit. To obtain the midpoint we divide the interval size by 2 and add this value to the lower real limit.

$$\frac{10}{2} = 5$$
$$289.5 + 5.0 = 294.5$$

So the midpoint of the interval is 294.5. Box 2.3 summarizes the steps to finding the midpoint.

Students sometimes get confused by the upper and lower real limits. A common error is to look at the interval 290–299 and think of it as representing 9 units ($299 - 290 = 9$). The mistake is in not considering the extra half-unit between the upper value and the upper real limit and between the lower value and the lower real limit.

Of course, intervals do not have to be in terms of integers. You can make them as small or as large as you like. Consider a grouped frequency distribution in which the interval is in hundredths or thousandths. The upper and

• BOX 2.3

Finding the Midpoint for a Class Interval

The midpoint of an interval can be found using the following formula:

$$MP = LL + (UL - LL)/2. \qquad (2.1)$$

where MP is the midpoint, LL is the lower real limit, and UL is the upper real limit.

As an example a class interval is shown in the graph below. Although the interval goes from 12 to 14, the lower real limit is 11.5 and the upper real limit is 14.5.

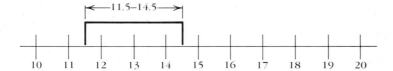

To find the midpoint perform the steps listed below.

STEPS

1. Identify the lower real limit (LL) and upper real limit (UL).

   LL = 11.5
   UL = 14.5

2. Find (UL − LL)/2.

   $$\begin{aligned}(UL - LL)/2 &= (14.5 - 11.5)/2 \\ &= 3/2 \\ &= 1.5\end{aligned}$$

3. Add the result of Step 2 to the lower real limit to obtain the midpoint (MP).

   11.5 + 1.5 = 13.

   Therefore, the midpoint is 13.

lower real limits are marked by the values halfway between these intervals. For example, if the scale is in tenths the intervals might be .10–.19, .20–.29, .30–.39, and so on. For the interval .20–.29 the lower real limit would be .195 and the upper real limit would be .295.

The size of an interval for grouped frequency data is an important choice that the statistician must make. Whenever data are grouped some information is ignored. In other words, by grouping information we decide to ignore the distribution within any interval. If, for example, there are 12 cases in the interval .20–.29, we treat each of those cases as though they were the same. If very large class intervals are used, and there are many cases within each, some important information may be ignored. The other extreme is when too many intervals are used. For example, if the class interval is extremely small (say, 1/1000) each case might fall in a different interval, and some intervals would not be used at all. Under these circumstances examining the frequency distribution may not be particularly informative.

In summary, the width of the interval represents a tradeoff between creating a meaningful visual display and ignoring some of the information. With very large intervals we cannot retrieve some important information about the distribution of scores within the intervals. With very small intervals the visual display may not lead to obvious interpretations. In practice most statisticians prefer to divide the range of scores into from 7 to 16 intervals. This decision is arbitrary and can vary from situation to situation.

Frequency Distributions for Nominal Scales

Frequency distributions involving raw scores and class intervals are useful when the scores are measured at at least an ordinal level. However, we often want to make frequency distributions to tally the number of times observations are made within nominal categories. For example, a university may wish to tally the number of students enrolled in the different majors. A major forms only a nominal scale of measurement. Nevertheless, a frequency distribution can be constructed.

When using frequency distributions, we often represent the frequencies as a percentage of the total number of cases rather than listing the raw frequencies. This often helps summarize large amounts of data. Table 2.3 shows the percentage of all murders in which different types of weapons were used. The frequency of murder by handgun is by far greater than that in any other category. Note that the raw numbers of murders are not tallied in each category. The frequencies are proportions rather than raw numbers, and we refer to this as a **relative frequency distribution.**

■ Cumulative Frequency Distributions

Sometimes we want to find the number of cases that fall below a particular score in a frequency distribution. To obtain this information easily, we use a **cumulative frequency distribution.** To obtain a cumulative frequency distribution we add a column labeled *cf* to the frequency distribution. Then, absolute frequencies are summed from the bottom of the column upward. For instance, Table 2.4 repeats the batting average data shown earlier in Table 2.2.

Table 2.3

Frequency data for type of weapon used in 1979 murders

WEAPON	RELATIVE FREQUENCY
Handgun	50%
Rifle	5%
Shotgun	8%
Cutting or stabbing	19%
Other weapon (club, poison, etc.)	12%
Personal weapon (hands, fists, feet, etc.)	6%

Source: 1979 FBI Uniform Crime Report.

Table 2.4

Baseball batting averages showing absolute, cumulative, relative, and cumulative relative frequencies

CLASS INTERVAL	ABSOLUTE f	CUMULATIVE cf	RELATIVE f *	CUMULATIVE RELATIVE f
320–329	2	89	.02	1.00
310–319	1	87	.01	.98
300–309	7	86	.08	.97
290–299	10	79	.11	.89
280–289	11	69	.12	.78
270–279	16	58	.18	.65
260–269	15	42	.17	.47
250–259	11	27	.12	.30
240–249	8	16	.09	.18
230–239	3	8	.03	.09
220–229	2	5	.02	.06
210–219	0	3	.00	.03
200–209	3	3	.03	.03

*Relative frequencies do not add to 1.0 because of rounding errors.

However, the cumulative frequency column has been added showing that there were three players batting 219 or less, five players batting 229 or less, eight players batting 239 or less, and so on. The cumulative frequency at the top of the column, 89, represents the total number of players whose batting averages fell between 200.5 and 329.5.

Graphs

By this time in your education you have probably seen many graphs and charts. Economic trends are shown in daily newspapers, and news magazines and television programs often present data graphically. In psychology, health science, or virtually any other field you choose, it will be important to understand graphical presentations and to be able to present data in the form of graphs. The visual display of quantitative information is not particularly new. As Biberman (1980) notes Shakespeare stated in *Henry VI* (Act IV): "Our forefathers had no other books but the score and the tally." Yet it was a long time before graphs found their way into common use. There were other graphic systems, including mechanical and architectural drawing, and musicians, alchemists, and others used scientific notation in drawings. By 1637 Descartes had introduced the Cartesian coordinate system, which forms the basis for modern graphs. The major transition to the modern use of graphics in science appeared in a book entitled *Political Atlas,* which was published by Playfair in 1786. Playfair introduced the bar chart, the histogram, and the pie diagram (Wainer and Thissen, 1981).

■ The Basics

Most of the graphs you use are two-dimensional, that is, they have two axes like those shown in Fig. 2.1. The axes are represented by two lines that are perpendicular to one another. Students sometimes get confused about the names of these axes because they are referred to by different names on different occasions. It will be important for you to learn all of the alternative names for each axis to avoid this common source of confusion. The horizontal axis is usually called the x-axis, although it is also known as the *abscissa*. Although not technically correct, the x-axis is often called the independent variable because values of the independent variable are often presented along the x-axis.

Similarly, the vertical axis goes by a variety of names. It is most often called the y-axis, although it is also known as the *ordinate*. Again although not technically correct, some researchers refer to the vertical axis as the dependent variable because in some studies the values of the dependent variable are presented along the y-axis.

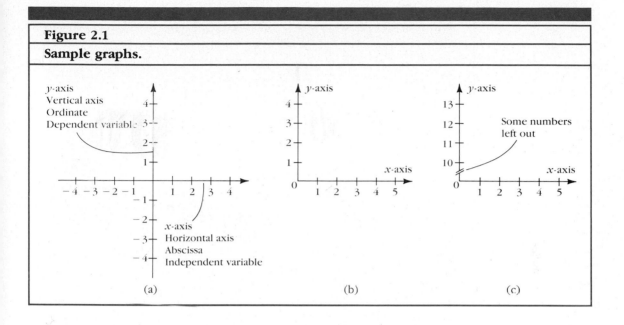

Figure 2.1

Sample graphs.

(a)   (b)   (c)

The graph shown in Fig. 2.1(a) includes both negative and positive values along each axis. Zero is in the center of each axis, and the graph looks like a large "+" sign. Students sometimes get confused because the graphs they see do not look like this. Instead, they look more like Fig. 2.1(b). Here the negative values have been eliminated along each axis, and the graph looks like half of a rectangle. Researchers usually present graphs this way even if they include negative values.

Another thing to look out for in a graph is whether there are break signs such as those shown in Fig. 2.1(c). These two lines suggest that some values have been left out along the axis. Here the y-axis increases in single unit increments from 10 through 13; however, at the bottom of the y-axis the graph goes from 0 to 10 without including the numbers in between. The break sign indicates that these numbers have been left out.

■ The Histogram

The histogram, or bar graph, is one of the most common ways to present numerical data. Most histograms are simply pictorial representations of frequency distributions. The score or variable of interest is plotted along the x-axis, and the frequency of the score or variable is represented by a vertical bar that extends up the y-axis.

Figure 2.2 is an example of a histogram for the data in Table 2.3. Handguns were used in 50% of the murders, so a vertical bar extends from the x-axis

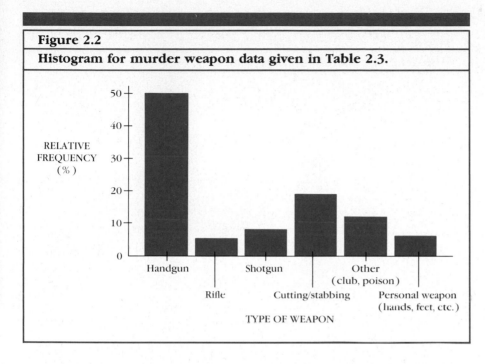

Figure 2.2

Histogram for murder weapon data given in Table 2.3.

above handgun to the point adjacent to 50% on the y-axis. Rifles were used in 5% of the murders, so a vertical bar above rifle extends from 0 to 5% on the y-axis, and so on. Note that in Fig. 2.2 the vertical bars do not touch one another. This is because the scale for the x-axis is nominal.

Figure 2.3 shows a histogram for the batting average data in Table 2.2. The class intervals for batting averages are presented along the x-axis. Vertical bars above each interval extend up the y-axis to show how many players were in each range. In contrast to Fig. 2.2, the bars in Fig. 2.3 touch one another. This is because the values along the x-axis are scaled and form a continuum.

It is also possible to build a histogram from a continuous frequency distribution. An example is shown in Fig. 2.4. The figure presents the continuous frequency distribution of batting averages (from Table 2.4) in histogram form. Instead of using raw frequencies (as in Fig. 2.3) Fig. 2.4 shows relative frequencies. Notice that as you move from left to right along the x-axis the height of the vertical bars becomes larger. The bar at the far right informs us that 100% of the players obtained batting averages of 329 or below. That is, no player has a batting average greater than 329. Each other bar in the graph tells us the percentage of players obtaining batting averages in that interval or below.

■ Frequency Polygon

Another way of looking at the same data is to present them in the form of a *frequency polygon*. The frequency polygon is very similar to a frequency his-

Figure 2.3

Histogram for batting average data.

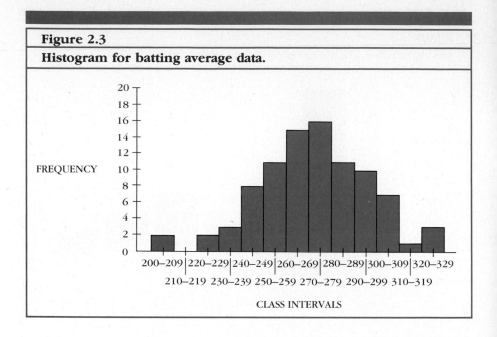

Figure 2.4

Histogram for continuous frequency distribution of batting average data.

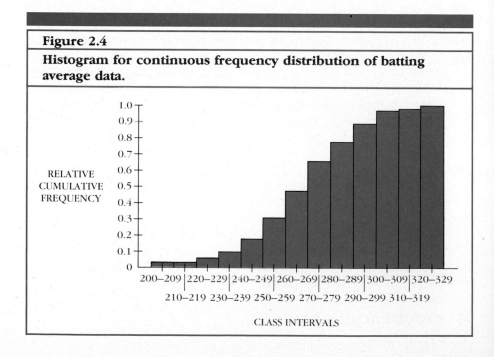

Figure 2.5

Frequency polygon for baseball data.

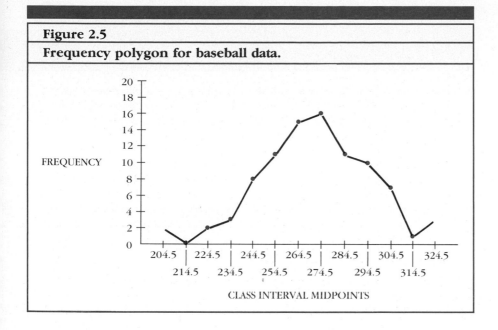

CLASS INTERVAL MIDPOINTS

togram except that points are used instead of bars. Lines are then drawn connecting the frequencies.

To create a frequency polygon for the batting average data we first obtain the midpoint for each class interval. The midpoints then become the x-axis values. Next we enter a dot above each midpoint corresponding to the y-axis value representing the frequency. Finally we connect the dots. Figure 2.5 is an example of the frequency polygon.

● BOX 2.4

"These Data"

The word data is Latin. Latin words ending with "a" are usually plural. Thus, data is the plural form. It is appropriate to refer to data as "they" or "these." Common phrases are "these data," and "the data are." The expression "the data is" is grammatically incorrect because of noun-verb disagreement; however, some scientists and dictionaries use data as a singular noun.

The singular for "data" is "datum." A single numerical score is referred to as a datum while an aggregation of scores is referred to as data.

Don't Be Fooled

People often try to deliberately mislead us by using statistical information. One of the purposes of your training in statistics is to learn how *not* to be misled. A common tactic is to take observations or scores that are very similar and try to make them look different in a histogram.

Here is how it is done. The person trying to mislead you usually puts breaks in the histogram and collapses the scores. An example is shown in Fig. 2.6. Figure 2.6(a) shows the average team batting averages for the six teams in the National League West at mid-season in 1985. It appears that they are very different because the vertical bars above each team reach very different heights. But the scale has been collapsed so that the lowest point on the scale is 235 and the highest is 260. Figure 2.6(b) places these same data in context by using the full range of batting averages. Notice that the averages for different teams now look quite similar.

You should always be very careful when interpreting a histogram. First look to see if breaks have been put in. Breaks are two parallel lines on the axis indicating the scale has been collapsed (see the figure below). Then, try to determine if differences are really meaningful. In later chapters you learn how to use statistical decision methods that will help you decide whether differences you see or observe are statistically meaningful.

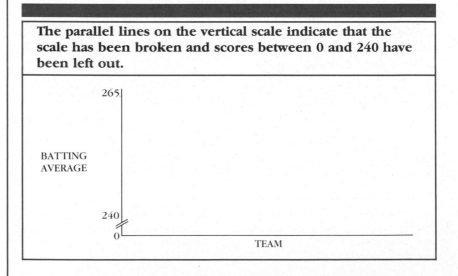

The parallel lines on the vertical scale indicate that the scale has been broken and scores between 0 and 240 have been left out.

Figure 2.6

(a) Example of a scale that has been collapsed; (b) example of a scale that has not been collapsed.

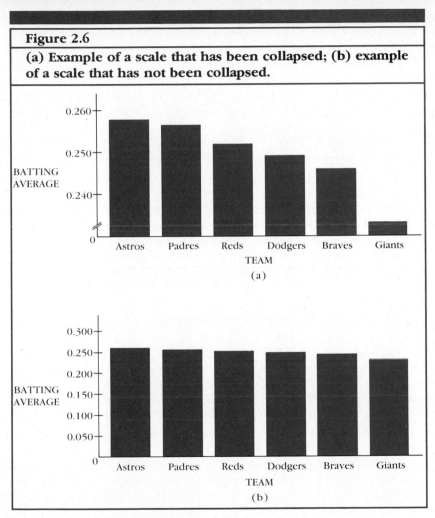

Source: Data from *Los Angeles Times,* July 14, 1985.

Summary

In this chapter we have taken some of the first steps toward making sense of numerical information. To understand quantitative information we must know the nature of the data. Measurements are represented by one of four types of scales. The nominal scale is a system that assigns numbers to categories. Nominal scales do not have properties that allow them to be used meaningfully in arithmetic operations. An ordinal scale is a scale that has the property of magnitude, allows us to rank objects, but that does not have the property of equal intervals or absolute zero. An interval scale can describe the distance between objects because it has the property of equal intervals in addition to the property of magnitude. An interval scale does not have the property of absolute zero. The scale that has the properties of absolute zero, equal interval, and magnitude is called the ratio scale. The ratio scale can be used in any mathematical operation.

Scores by themselves are often difficult to interpret. To make more sense of them we usually examine individual scores in relation to other scores. This requires a distribution of scores. There are many ways to create distributions. In a raw frequency distribution, we tally the number of times each score occurs. We can also create frequency distributions for groups of scores. For example, the number of times observations are made within defined intervals can be used to create frequency distributions. The ranges of scores that group the variables are called class intervals. In addition to the specified upper and lower limits of a class interval, each interval has an upper real limit and a lower real limit. The upper and lower real limits extend halfway to the next interval on each side. A relative frequency distribution shows the proportion of cases at each score or interval. A cumulative relative frequency distribution shows the proportion of cases at or below each score or interval.

There are different methods for graphing frequency distributions. One of the most common methods is called the frequency histogram, which is also known as the bar graph. In the frequency histogram vertical bars extend above each point or interval to show their frequency. A similar method is called the frequency polygon. In the frequency polygon points represent the frequencies of observations for different scores or categories. Lines are used to connect the points.

Distributions are very important in the field of statistics. In later chapters methods for relating theoretical and observed distribution scores are presented.

Key Terms

1. **Magnitude**   A property of "moreness." A scale has the property of magnitude if we can say that one instance of the attribute represents more, less, or equal amounts than another instance.
2. **Equal Intervals**   A scale has the property of equal intervals if the meaning of the difference between two points is uniform along the entire scale.
3. **Absolute Zero**   A scale has the property of absolute zero if there is a meaningful point at which nothing at all of the property exists.
4. **Nominal Scale**   A scale that does not have the property of magnitude, equal interval, or absolute zero.
5. **Ordinal Scale**   A scale with the property of magnitude but not the property of equal interval or absolute zero.
6. **Interval Scale**   A scale with the properties of magnitude and equal interval but not the property of absolute zero.
7. **Ratio Scale**   A scale that has the properties of magnitude, equal interval, and absolute zero.
8. **Continuous Variable**   A variable that may take on any value within a defined range. Continuous variables can be reported in any measurement unit, however small.
9. **Discrete Variable**   A variable that is not continuous. It is expressed in a well-defined measurement unit such as an integer or specific fraction.
10. **Frequency Distribution**   A simple way of explaining and summarizing numerical information based on frequency of occurrence of events in the categories of a nominal scale.
11. **Class Interval**   A portion of a measurement scale containing more than one possible value. The class interval is defined by the upper real limit and the lower real limit. The upper real limit extends halfway to the lower real limit of the interval above. Similarly, the lower real limit extends halfway down to the upper real limit of the interval below.
12. **Relative Frequency Distribution**   A frequency distribution in which the frequencies are proportions rather than raw numbers.
13. **Cumulative Frequency Distribution**   A relative frequency distribution in which the sum of the frequencies falling within and below each interval is reported.
14. **Abscissa**   The horizontal axis on a graph.
15. **Ordinate**   The vertical axis on a graph.

16. **Histogram**   A graph of a frequency distribution in which each frequency is represented by a bar.
17. **Frequency Polygon**   A graph similar to a frequency histogram except that frequencies are represented by points and the points are connected by a line. Points represent the midpoints of each class interval in the frequency distribution.

Exercises

Multiple Choice

1. A scale has the property of magnitude but not the property of equal interval or absolute zero. What type of scale is it?
    a. nominal
    b. ordinal
    c. interval
    d. ratio

2. A scale has the property of magnitude and equal interval but not the property of absolute zero. What type of scale is it?
    a. nominal
    b. ordinal
    c. interval
    d. ratio

3. A physical education class participates in a jump and reach test. The number of inches students were able to jump are presented in the table below. Make a frequency distribution for these jumping scores.

Student Initial	Score
PT	4
WT	1
HH	2
BH	3
WA	1
AW	2
BT	3
TT	1
ET	0
CA	5
CD	2
JB	1

4. Make a frequency histogram for these scores.

5. Make a frequency polygon for these scores.

6. Make a cumulative frequency distribution for these scores.

7. Table 2.4 shows baseball batting averages. What is the lower real limit for the interval 240–249? What is the upper real limit for the interval 210–219?

3

Descriptive Statistics

The patient sat anxiously as the doctor returned to the examining room holding the file folder. "I have your test results here." "What are they?" the patient intensely questioned. "Well your hematocrit is 46%, your white count is 8,000, your PO_2 is 89, and your PcO_2 is 21."

These numbers are essentially meaningless to the patient. Are they unusual? Are they anything to be concerned about? To answer these questions we have to put the scores in some context. One of the most important questions is whether these values are average, below average, above average, or abnormal in any way. In this chapter we study methods for describing and summarizing distributions of scores.

There is usually too much information to attend to in a distribution. One of the purposes of statistics is to summarize this information. Two different summaries are measures of central tendency and measures of variability. In this chapter central tendency and variability summary statistics are presented.

Measures of Central Tendency

What is a typical score? In statistics there are at least three different ways to determine if a score is typical. The typical score usually occurs in the center of distribution, and indexes of typicalness are usually called measures of central tendency. The three measures of central tendency to be explored in this chapter are the mean, the median, and the mode. The mean is the arithmetic average, the median is the point representing the 50th percentile in the distribution, and the mode is the most common score. Sometimes each of these measures is the same. Yet on other occasions the mean, the median, and the mode can be different. Thus, it will be important for you to know each type of central tendency and to distinguish among them.

■ The Mean

Among the measures of central tendency the mean is most commonly used in statistics. The mean is simply the arithmetic average. An example of the calculation of the mean is shown in Box 3.1. The table presents the murder rates (number of homicides per 10,000 for 10 states). A *variable* is a quantity that may take on different values. Murder rate is a variable because the number of persons killed throughout the country is not always the same. We represent the variable as X. Then the murder rates in the different states are arbitrarily denoted $X_1, X_2, X_3, \ldots, X_N$. X_N is for the last score. In the example there are ten states, so $X_N = X_{10}$. We can also say that N, the number of cases, $= 10$. To obtain the mean we use the formula

$$\bar{X} = \frac{\Sigma X}{N}.$$

(3.1)

The mean, $\bar{X}$, is pronounced "X bar" to describe the bar on top of the X. The symbol Σ is the Greek capital letter sigma that instructs you to sum. In this case ΣX means you sum the X's. Then you divide by the number of cases N.

There are three steps to obtaining the mean. First you must find the sum of the X values, or

$$\sum X.$$

This is accomplished by adding the murder rates for the ten states, that is, we add $15 + 10 + 2 + \cdots + 5$. In this example the sum equals 66.

Next we find N simply by counting the number of scores. In this example $N = 10$.

Now the mean is obtained by dividing the sum of the X values by N, or

$$\bar{X} = \frac{\Sigma X}{N} = \frac{66}{10} = 6.6.$$

Thus, the results of this calculation show that the mean, or average, murder rate is 6.6.

The mean has many properties. The most important of these is that the sum of the deviations around the mean always equals zero. In other words, the sum of the absolute value of the scores above the mean always equals the sum of the absolute value of deviations below the mean. But because the sum below the mean is negative, adding the two sums gives zero. Box 3.3 shows an

Calculation of the Mean:
Example of Murder Rates* in 10 States

STATE	VARIABLE X	MURDER RATE
Alabama	X_1	15
California	X_2	10
Iowa	X_3	2
Mississippi	X_4	12
New Hampshire	X_5	3
Ohio	X_6	7
Oregon	X_7	4
Pennsylvania	X_8	6
South Dakota	X_9	2
Vermont	X_{10}	5
		$\Sigma X = \overline{66}$

STEPS

1. Find the sum.

    $$\Sigma X = 15 + 10 + 2 + 12$$
    $$+ 3 + 7 + 4 + 6$$
    $$+ 2 + 5$$
    $$= 66$$

2. Find N by counting the number of scores.

    $$N = 10$$

3. Calculate the mean by dividing the result of Step 1 by the result of Step 2.

    $$\overline{X} = \frac{\Sigma X}{N} = \frac{66}{10} = 6.6$$
    $$\overline{X} = 6.6$$

*Rounded to the nearest whole number. Rates are homicides per 10,000.
Source: 1978 FBI Statistics.

● BOX 3.2

Common Symbols

Some symbols will be used throughout the entire book, and it is important that you understand and learn to recognize them.

$\overline{X}$ is the mean. It is pronounced X bar.

Σ is the summation sign. It means sum or add scores together, and is the Greek letter sigma.

X is a variable that takes on different values.

● BOX 3.3

The Sum of the Deviations Around the Mean Is Zero

STATE	VARIABLE	X	$\overline{X}$	$x = X - \overline{X}$
California	X_2	10	6.6	3.4
Iowa	X_3	2	6.6	−4.6
Mississippi	X_4	12	6.6	5.4
New Hampshire	X_5	3	6.6	−3.6
Ohio	X_6	7	6.6	.4
Oregon	X_7	4	6.6	−2.6
Pennsylvania	X_8	6	6.6	−.6
South Dakota	X_9	2	6.6	−4.6
Vermont	X_{10}	5	6.6	−1.6
				$\Sigma(X - \overline{X}) = 0$

STEPS

1. Find the deviation, X, of each score around the mean. This is done by subtracting the mean from each score.

   $x = X - \overline{X}$

2. Sum the deviations. You should get 0.

example of this property using the data from Box 3.1. The importance of this will become obvious as we discuss other measures of central tendency. You should work through this example and others to understand this property.

■ The Median

Another way of describing the center of distributions of scores is the median. The **median** is the 50th percentile, or the point in the distribution where half of the cases fall above and the other half fall below.

The median is used less frequently than the mean in statistical analysis. To calculate the median, scores must be rank ordered. The method differs slightly if there is an even or an odd number of scores. Let's take an odd number of scores first.

An example of the calculation of the median for an odd number of scores is presented in Box 3.4. The scores are batting averages for 11 National League baseball players. The first step is to make a list of the scores in descending order. In other words, the highest score is placed at the top of the list, the second highest score is listed next, and so on, so that the lowest score is listed at

● BOX 3.4

Calculation of the Median for Odd Number of Scores Using Batting Averages for 11 National League Players

STEPS

1. Make a list of scores placing the highest on the top and ranking others in descending order.

2. Find the depth of median.

$$Md_m = \frac{N + 1}{2}$$

$$= \frac{11 + 1}{2}$$

$$= 6$$

The depth of median is 6.

3. Count down 6 scores from the top or count up 6 scores from the bottom. You should stop at 275. This is the median.

PLAYER	BATTING AVERAGE	
Herr	330	
Krechicki	300	
Gwynn	295	
Murphy	288	
Ramirez	279	
Brock	275	← median
Carter	268	
Kennedy	264	
Bailey	258	
Dernier	253	
Milner	241	

Data source: Los Angeles Times, July 28, 1985.

the bottom. The second step is to find the depth of median (Tukey, 1978) using the formula

$$Md_m = \frac{N + 1}{2}$$

In Box 3.4 there are eleven scores. Therefore, $N + 1 = 12$, and $12/2 = 6$. This tells us that the depth of median is 6. Next, we use this information to count in the depth. Counting in can begin at either the bottom or the top. In the example, if you count down 6 scores from the top or up 6 scores from the bottom, you should arrive in either case at the score .275. This is the median.

Calculation of the median for an even number of scores is more difficult because the depth of median may fall between two scores. An example of the calculation of the median for an even number of scores is presented in Box 3.5. This example uses the homicide data that were used in the calculation of the mean (see Box 3.1). To find the median for an even number of scores, the scores must be listed in descending order. Next, you must calculate the depth of median using the formula

$$Md_m = \frac{N + 1}{2} \tag{3.2}$$

(The symbol Md_m refers to the depth of median. The median will be denoted as Md.) In Box 3.5 there are ten scores, so the depth of median is $(10 + 1)/2$, which is 5.5. This tells us that the median is between the 5th and 6th score from either the bottom or the top of the distribution. If we count in 5.5 scores from either the bottom or the top, we are between the number 6 and the number 5. To find the median we add these two scores together and divide by 2. In other words, the median is the average of the two scores that the depth of median falls between. So using the scores 6 and 5 we obtain $(6 + 5)/2 = 11/2 = 5.5$. Thus, the median is 5.5.*

■ The Mode

In addition to the mean and the median, the mode is sometimes referred to as a measure of central tendency. However, the mode can give very different information than the mean or median for some distributions of scores.

The mode is the score that occurs most frequently in the distribution. In Chapter 2, data were presented on batting averages for National League players. The histogram in Fig. 2.3 is reproduced as Fig. 3.1. As the histogram shows, more players were in the class interval 270–279 than in any other interval. Thus, we would refer to this interval (or the midpoint of this interval) as the

---

*Some textbooks recommend interpolation procedures when there is duplication of scores near the median. In practice, however, this is not commonly done, and interpolation methods will not be presented here. The interpolation methods are technically more accurate, and interested readers may refer to Hays, 1973.

• BOX 3.5

Calculation of the Median for an Even Number of Scores Using Homicide Rate Data

STEPS

1. Make a list of scores placing the highest score at the top of the list and ranking others in descending order.
2. Calculate the depth of median.

$$Md_m = \frac{N + 1}{2} = \frac{10 + 1}{2}$$

$$= 5.5$$

3. Since the depth of median falls halfway between two scores, the median is halfway between them. You can obtain this value by summing the scores and dividing by 2. In this example $(6 + 5)/2 = 11/2 = 5.5$.

STATE	HOMICIDE RATE
Alabama	15
Mississippi	12
California	10
Ohio	7
Pennsylvania	6
	5.5 ← median
Vermont	5
Oregon	4
New Hampshire	3
South Dakota	2
Iowa	2

Figure 3.1
Histogram for batting average data.

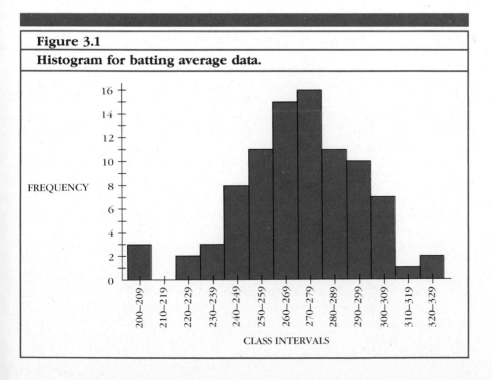

FREQUENCY

CLASS INTERVALS

mode because it is the most common. In this example the mode is close to the mean and the median.

There are other cases in which the mode can be different from the mean and the median. Consider the data on homicide rates in Boxes 3.1 and 3.5. In this example, the most frequent homicide rate was 2 because it occurred more than any other score. This was also the lowest homicide rate observed in any of the states in the sample. Thus, in the example of homicide rates, the mode is quite different from either the mean or the median. Finally, consider the example of the batting averages for the 11 players presented in Box 3.4. In this example, each player had a different batting average. In other words, no score occurred more frequently than any other score; each score occurred just once. Thus, there is no mode in this example and it cannot be compared to the mean or the median.

There are many cases in which the mode is useful. The most common of these is when a nominal scale is used. For instance, if we want to tabulate the number of people enrolled in different majors in college, the mode is quite appropriate. We might find that more students major in Business than in any other major, making Business the modal major. Here it would make no sense to calculate a mean or a median. Another example is the weapon used to commit murder as shown in Fig. 2.2. Handgun was the most common murder weapon, so handgun is the mode for this distribution. Again it would make no sense to calculate a mean or a median for this distribution of scores.

■ Similarities and Differences Among the Mean, Median, and Mode

The mean, median, and mode are all considered measures of central tendency, yet as we have seen their values are not always the same. The mean, median, and mode are the same when the distribution of scores is normal. The **normal distribution** is a theoretical distribution of scores that is symmetrical. In a normal distribution the mean, median, and mode have the same value, and half the scores fall below that value and half fall above. We talk about the normal distribution and characteristics of nonnormal distribution in Chapter 4. In that chapter, the relationship among the normal distribution, the mean, and the median are discussed. For now you should remember that in the theoretical normal distribution, the mean, median, and mode have exactly the same values. In most real data sets, they will differ.

With the exception of analyses involving nominal data, the mode is rarely if ever used in statistical analysis. As we have seen in the example of homicide rates (Box 3.5) the mode can be quite different from the mean or median and is rarely the very best way to summarize a group of scores.

We usually choose between the mean and the median as a measure of central tendency. The major consideration here is how much weight to give to extreme scores. The mean takes into account each score in the distribution; the median finds only the halfway point.

Table 3.1 shows how the mean and median might have very different values. The table lists the weights of two groups of college women. There are five women in each group, but four of them are in both groups (Sally, Sue, Leslie, and Dana). The fifth member is Allison in Group 1 and Bertha in Group 2. First look at Group 1. Here the mean and the median are exactly the same—105 in each case. Now look at Group 2. Four of the five women in this group are exactly the same as those in Group 1; however, Allison, who weighs 115 pounds, is replaced by Bertha, who weighs 275 pounds. Does the substitution of Bertha for Allison affect the average weight? The answer depends on which measure of central tendency is used. Substituting Bertha for Allison has no effect on the median. This is because the median is the middle point in the distribution, and the substitution is made at one extreme. On the other hand the mean considers each value and is affected by each score in the distribution. Thus, the mean is affected by the substitution of Bertha. With her 275 pounds in the group, the mean increases considerably.

This example may have led you to believe that the mean is the more appropriate index of central tendency. Indeed there are many advantages to the mean. These include the following:

The mean considers all scores in the distribution.
The mean is often easier to calculate.
The mean is used most often in formal statistical analysis.

We use the mean in nearly every chapter of this book.

Despite the advantages of the mean, there are also some advantages to using the median. One of my colleagues who favors the median argues that the mean gives too much weight to "oddballs" in a distribution, that is, that extreme scores have a bigger impact on the mean than less extreme scores. If there are very large or very small scores in a distribution, the mean moves in

Table 3.1

Mean and median for weights (in pounds) of two groups of college women

GROUP 1		GROUP 2	
Woman	Weight	Woman	Weight
Sally	95	Sally	95
Sue	100	Sue	100
Leslie	105	Leslie	105
Dana	110	Dana	110
Allison	115	Bertha	275
Mean $\bar{X} = 105$		Mean $\bar{X} = 137$	
Median Md $= 105$		Median Md $= 105$	

the direction of these scores more than does the median. For instance, if we have a distribution with scores 1, 2, 3, and 1 million, the score 1 million has a much stronger influence on the mean than do the other scores. Here the mean is 250,002. The median is 2.5. Most scores in the distribution are less than 4, and the very large score has an unmerited impact on the measure of central tendency.

In summary, the median is useful when we do not want scores at the extreme of a distribution to have a strong impact. In addition there are certain occasions in which the extreme of a distribution may not be measured accurately. For instance, consider an easy test on which 75% of the students get a perfect score. This might be a small class of four students in which the scores are 100, 100, 100, and 35. The median is the number of items possible, since more than half of the class got this score. A very easy test may have a ceiling effect that does not show the true ability of some test-takers. A ceiling effect occurs when the test was too easy to measure the true ability of the best students. Thus, if some scores stack up at the extreme, the median may be more accurate than the mean. If the high scores had not been bounded by the highest obtainable score, the mean may actually have been higher.

In the normal distribution, the mean, median, and mode are exactly the same. However, not all distributions of scores have a normal or bell-shaped appearance. The highest point in a distribution of scores is called the modal peak. A distribution with the modal peak off to one side or the other is described as skewed. The word skew literally means "slanted."

The direction of skew is determined by the location of the tail or flat area of the distribution. Positive skewness occurs when the tail goes off to the right of the distribution. Negative skewness occurs when the tail or low point is on

Figure 3.2

Examples of normal distribution, positive skewness, and negative skewness.

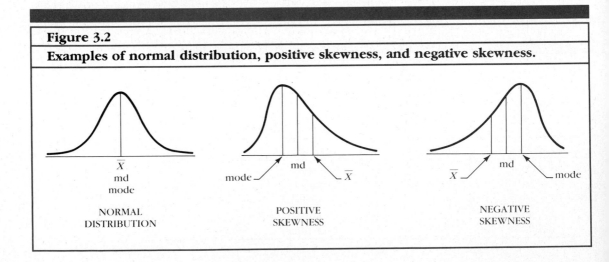

the left side of the distribution. Figure 3.2 illustrates the normal distribution, a distribution that has positive skewness, and one that has negative skewness.

The mode is the most frequent score in a distribution. In a skewed distribution, the mode remains at the peak. The mean and the median shift away from the mode in the direction of the skewness. The mean moves furthest in the direction of the skewness, and the median typically falls between the mean and the mode. The relative positions of the mean, median, and mode in normal and skewed distributions are shown in Fig. 3.2.

Skewed distributions are examples of situations in which different measures of central tendency give different information. In contrast to the normal distribution in which all measures of central tendency are the same, the mean and the median shift in the direction of the tail of a skewed distribution. The mean shifts furthest in the direction of the skewness.

Measures of Variability

Measures of central tendency such as the mean and median are used to summarize information. They are important because they tell something about the average score in a distribution. Knowing the average score, however, does not give us all the information that we need about a group of scores. As an illustration look at the following two sets of numbers.

SET 1	SET 2
6	4
6	6
6	8
6	3
6	6
6	9

Calculate the mean for the first set. Did you get 6? You should have. Now calculate the mean for the second set. If you calculated correctly you should get 6 again. In other words, the two distributions of scores that look very different give you exactly the same mean. Yet we would not want to say that the two distributions are equivalent. In addition to measures of central tendency, we need other indicators that summarize characterisitics of the scores. Some of the most common indicators are the range, the variance, and the standard deviation.

■ The Range

The simplest method for describing variability is the **range**. The range is simply the difference between the highest score and the lowest score. In the last example the range for Set 1 is 0 since the highest score is 6 and lowest score is 6.

$$\text{Range} = \text{Highest score} - \text{Lowest score}$$
$$= 6 - 6 \qquad\qquad (3.3)$$
$$= 0$$

For Set 2 the range is

$$9 - 3 = 6.$$

Although the range is simple to understand and easy to calculate, it has very few uses in statistical analysis. Measures of variability summarize information about distributions of scores. The range provides very little information because it uses information only from two scores at the extreme. However, classroom teachers often find it useful to show the lowest and highest score on an exam, and the range does tell how good or bad a score can be. The range is also commonly noted in record books. People are always interested in knowing the temperatures for the hottest and coldest days on record, who the tallest and shortest human is, and so forth. Yet with few exceptions the range has little practical value in data analysis.

Another statistic known as the interquartile range describes the interval of scores bounded by the 25th and 75th percentile. In other words, the interquartile range is bounded by the range of scores that represents the middle 50 percent of the distribution. The interquartile range is used considerably in statistical analysis and in the field of test and measurement. It is more useful than the range because it provides some information about the characteristics of the distribution. Methods for calculating the interquartile range are given when percentiles are discussed in Chapter 4.

■ The Variance and the Standard Deviation

In contrast to ranges, which are used infrequently in statistical analysis, the variance and the standard deviation are used frequently. As we noted in the last section the mean is a reasonably good descriptor of central tendency. Given the advantages of the mean we want to use a meanlike measure to express variability. In addition to stating the mean we may want to describe the average deviation around the mean. In other words, we could sum the deviations around the mean and divide by the number of observations. However, this approach will not be fruitful. Recall Box 3.3 where the deviations around the mean were calculated. That box gives an example in which the deviations around the mean summed to zero, and we noted that this was not an unusual case. In fact, you should get zero each time you sum the deviations around the mean. Before going further make up several small distributions of scores. Then calculate the sum of the deviations around the mean. The sum should always be zero.

An example of the calculation of the mean of deviations around the mean is presented in Box 3.6. In this example you are asked to demonstrate to yourself that the mean of deviations around the mean is zero. To understand Box 3.6 you might want to look at Box 3.7, which cautions you that "big" X and "small" x have different meanings. Statisticians typically refer to an uppercase X as a value of a variable and lowercase x as the deviation between the value and the mean for the variable.

Demonstrate to yourself that the mean of deviations around the mean is zero. There are four steps to this demonstration. First make up a distribution of scores. I have used 2, 4, 6, 1, 4, 7. You should choose other numbers. Now find the deviation of each score around the mean. For example, the deviation of my first score, 2, around the mean, which in this example is 4, is

$$2 - 4 = -2.$$

The deviation of the second score, 4, around the mean is

$$4 - 4 = 0,$$

and so on. Next find the mean of the deviations around the mean. Remember that a lowercase x is the deviation of $X - \overline{X}$. So the mean of the deviations

• BOX 3.6

Mean of Deviations Around the Mean

Demonstrate to yourself that the mean of deviations around the mean is zero

STEPS

1. Make up a distribution of scores.

2. Find the mean.

$$\overline{X} = \frac{\Sigma X}{N}$$

3. Find the deviation of each score around the mean.

$$x = X - \overline{X}$$

4. Find the mean of the deviations around the mean.

$$\overline{X} = \frac{\Sigma x}{N}$$

AN EXAMPLE		
2		
4		
6		
1		
4		
7		
$\Sigma x = 24$		
$N = 6$		
$\overline{X} = 4$		
X	$\overline{X}$	x
2	4	−2
4	4	0
6	4	2
1	4	−3
4	4	0
7	4	3
	$\Sigma x = 0$	
	$N = 6$	
	$\overline{x} = 0$	

• BOX 3.7

Watch Out for the Big *X* and the Small *x*

Upper and lowercase letters do not always have the same meaning in statistics. One source of confusion for many students is the difference between the uppercase X that stands for a value of a variable, and the lowercase x that is the deviation between the variable and the mean of the distribution. Symbolically,

$$x = X - \overline{X}.$$

In words, lowercase x equals the deviation of a score around the mean.

around the mean is

$$\bar{X} = \frac{\Sigma x}{N},$$

where x is the deviation score, and Σx is the sum of the deviation scores. In my example, $\Sigma x = 0, N = 6$, so

$$\frac{\Sigma x}{N} = \frac{0}{6} = 0.$$

The result (0) should be exactly the same for your example. In fact in any example you choose, the mean of the deviations around the mean will be zero.

In summary it seems reasonable to use average deviation around the mean as an indicator of variability. Yet this is not feasible because the mean of the deviations around the mean always equals zero. We need some alternative as a measure of variance.

In exploring explanations for why the sum of the deviations around the mean equals zero, we find that the sum of deviations with "+" signs equals the sum of deviations with "−" signs. If we could get rid of the negative sign, then the sum of the deviations would not equal zero. One way to get rid of negative signs is to square all of the deviations around the mean. Then we can obtain the average *squared* deviation around the mean. This is known as the variance. The formula for the variance is

$$\sigma^2 = \frac{\Sigma(X - \bar{X})^2}{N}. \tag{3.4}$$

The variance is a very useful statistic and is commonly used in data analysis. However, to calculate the variance it was necessary to find the squared deviations around the mean rather than the simple deviation around the mean. Remember that the variance is the average *squared* deviation around the mean. Thus, when the variance is calculated, the result will always be in squared units. How can we get the result back into units that will make sense to us? The answer is by taking the square root of the variance. Since the deviations were originally squared, taking the square root of the variance places the variance in more familiar terms. We refer to the square root of the variance as the *standard deviation,* which is represented by the lowercase Greek letter sigma. The formula for the standard deviation is

$$\sigma = \sqrt{\frac{\Sigma(X - \bar{X})^2}{N}}. \tag{3.5}$$

In other words, the standard deviation is the square root of the variance, the average squared deviations around the mean. An example of the calculation of the standard deviation is shown in Box 3.8.

Some students have asked why we go through all this when we could seemingly take the mean of the absolute deviation around the mean, or

$$\frac{\Sigma|X - \bar{X}|}{N} = \frac{\Sigma|x|}{N}.$$

Remember that if we sum the deviations around the mean, we will always get 0 (see Box 3.6). Therefore, the only options are to sum the absolute deviations or the squared deviations of scores about the mean. The reason that we do not use the average absolute deviation is that the variance and standard deviation have certain properties that make them useful in formal operations. Knowing the standard deviation of a normal distribution (see Chapter 4) allows us to make precise statements about the distribution. The sum of these squared deviations is more easily manipulated in analyses and in proofs than is the sum of the absolute deviation. Although the standard deviation is not technically equal to the average deviation, it gives an approximation of how much scores deviate from the mean *on the average*. In other words, the results are similar with either approach.

The formulas presented above are what is known as *definitional formulas*. This means that they provide the definition of the variance and the standard deviation. However, it is often easier to use a mathematically equivalent formula to calculate the mean and the standard deviation. The computational formula for the standard deviation is

$$\sigma = \sqrt{\frac{\Sigma X^2 - \dfrac{(\Sigma X)^2}{N}}{N}}. \tag{3.6}$$

Box 3.8 uses a computational formula. The box presents eight steps in the calculation of the standard deviation. To begin you calculate the sum of X just as you did in calculating the mean. Then you square each score. For example, the first value, 15, is squared to obtain $15 \times 15 = 225$. In Step 3 you calculate the sum of the squares.

Next the result of Step 1 is squared to obtain $(\Sigma X)^2$. This is the step where students sometimes have problems. It is important that you not confuse $(\Sigma X)^2$ (Step 4) with ΣX^2 (Step 3). $(\Sigma X)^2$ has parentheses; ΣX^2 does not. To obtain $(\Sigma X)^2$, you must first sum the X's and then square the total. ΣX^2 is obtained by squaring each value first, then summing the squared values. In Step 5 $(\Sigma X)^2$ is divided by N, and the result is subtracted from the result of Step 3. When this value is divided by N, the result is the variance (Step 7). Finally the standard deviation is found by taking the square root of the variance (Step 8). You should work through this example and then try some of the problems in the workbook. In practice you may be calculating the standard deviation using an electronic calculator or a computer. Even so, working through these steps will give you a better understanding of the standard deviation.

Calculation of the Standard Deviation

Formula:
$$\sigma = \sqrt{\dfrac{\Sigma X^2 - \dfrac{(\Sigma X)^2}{N}}{N}}$$

$$\frac{612 - 435.6}{N} = \frac{176.4}{10} = 17.64 \text{ variance}$$

(3.6)

DATA
(from Homicide Rates in 10 States—see Box 3.1)

State	X	X^2
Alabama	15	225
California	10	100
Iowa	2	4
Mississippi	12	144
New Hampshire	3	9
Ohio	7	49
Oregon	4	16
Pennsylvania	6	36
South Dakota	2	4
Vermont	5	25
	$\Sigma X = 66$	$\Sigma X^2 = 612$

STEPS

1. Calculate ΣX.

   $\Sigma X = 15 + 10 + 2 + 12 + 3 + 7$
   $+ 4 + 6 + 2 + 5 = 66$

2. Calculate X^2 for each score. These values are shown in the second column in the table.

   $(15)^2 = 15 \times 15 = 225$
   $(10)^2 = 10 \times 10 = 100$
   $\vdots$
   $(5)^2 = 5 \times 5 = 25$

3. Calculate ΣX^2.

   $\Sigma X^2 = 225 + 100 + 4 + 144 + 9 + 49$
   $+ 16 + 36 + 4 + 25 = 612$

4. Find $(\Sigma X)^2$ by squaring the result of Step 1.

   $(\Sigma X)^2 = \Sigma X \times \Sigma X = 66 \times 66 = 4356$

   CAUTION: ΣX^2 and $(\Sigma X)^2$ are *not* the same thing! ΣX^2 is the sum of the individual squared values of X. $(\Sigma X)^2$ requires that you sum the individual values first, then *square the total.*

5. Divide $(\Sigma X)^2$ (the result of Step 4) by N.

   $\dfrac{(\Sigma X)^2}{N} = \dfrac{4356}{10} = 435.6$

6. Subtract the result of Step 5 from the result of Step 3.

   $612 - 435.6 = 176.4$

7. Divide the result of Step 6 by N.

   $\dfrac{176.4}{10} = 17.64$

   This is the variance.

8. Find the standard deviation by taking the square root of the variance.

   $\sigma = \sqrt{17.64} = 4.2$

Samples and Populations

Now we come to one of the most important concepts in the book: the distinction between a population and a sample. The methods we have used so far describe populations. A **population** is defined as the entire collection of a set of objects, people, events, etc., of interest in a particular context (Yaremko, Harari, Harrison, and Lynn, 1982). In other words, population refers to the collection of all items that we want to make generalizations about. For instance, in the example of batting averages for National League players, the population included all the players who had been at bat a sufficient number of times.

Actually, in statistics, we calculate means and standard deviations for populations less frequently than we calculate these values for samples. A **sample** is a subset of observations selected from a population. In Box 3.4 the mean batting average for a subgroup of National League players was calculated. In other words, a sample of players was selected from the population of National League players, and the mean was calculated for the sample.

In later chapters we review in detail the methods for estimating characteristics of the population on the basis of observations of a sample. Usually, it is not feasible for us to calculate the mean and standard deviation for the entire population. For example, if we want to determine heights of American men we would have to measure each man in America. However, we know that if we select a random and unbiased sample of American men we can estimate the mean of the population on the basis of the sample. The mean and the standard deviation of a sample that is drawn randomly from a population provide unbiased estimates of the population characteristics. Of course, using the sample mean and standard deviation may also lead to some error.

For reasons described in more detail in later chapters, we use different formulas for samples and for populations. For instance, the formula for calculating the standard deviation of a sample is slightly different from that for a population. The denominator in the radical, $N - 1$, is used for sample statistics instead of N, which is used for population statistics. Thus, the definitional formula for the standard deviation of a sample is

$$S = \sqrt{\frac{\Sigma(X - \bar{X})^2}{N - 1}}. \tag{3.7}$$

Similarly the computational formula for a sample is

$$S = \sqrt{\frac{\Sigma X^2 - \frac{(\Sigma X)^2}{N}}{N - 1}}. \tag{3.8}$$

Beware of Greek and Roman Letters

Greek letters such as σ usually describe population parameters while Roman letters such as S are used to describe sample statistics.

Sample	**Population**

$$S = \sqrt{\frac{\Sigma(X - \bar{X})^2}{N - 1}} \qquad (3.9) \qquad \sigma = \sqrt{\frac{\Sigma(X - \bar{X})^2}{N}} \qquad (3.10)$$

For the standard deviation, the Roman S tips you off that a sample standard deviation is being calculated and that $N - 1$ rather than N should be used. The Greek letter σ indicates that N should be used.

Also notice in the two formulas above that we have used the letter S instead of σ. This is characteristic in data analysis. Greek letters typically describe parameters or characteristics of populations, whereas Roman letters (the alphabet we use in English) describe sample statistics. Sample statistics (Roman letters) are used to estimate population parameters (Greek letters).

Summary

This chapter covers methods for summarizing distributions of scores. Three commonly used methods describing the central tendency of distributions are the mean, the median, and the mode. The mean is the arithmetic average score. The median is the point representing the 50th percentile. This is the point that divides the distribution into two halves with equal numbers of cases. The mode is the score within the distribution that occurs most frequently.

In addition to measures of central tendency, distributions are described according to their variability. The simplest index of variability is the range, which is the difference between the highest and lowest score. The variance is the average squared deviation around the mean. The standard deviation is the square root of the variance.

There are small differences between the formulas for the variance and the standard deviation for populations and for samples. Populations include every object characterized by the same definition. A sample is a subset of cases selected from the population. Measures of central tendency and variability are important building blocks for nearly all statistical methods.

Key Terms

1. **Mean**   The arithmetic average. The mean is obtained by summing scores across individual cases and dividing the sum by the number of cases.
2. **Median**   The point representing the 50th percentile in a distribution of scores, that is, the point that divides the distribution into upper and lower halves.
3. **Mode**   The score in a distribution which occurs with greatest frequency.
4. **Normal Distribution**   A theoretical distribution of scores in which the mean, median, and mode are the same. The other scores in the normal distribution are symmetrically distributed around these measures of central tendency.
5. **Range**   The difference between the highest and lowest score in a distribution. The range is calculated by subtracting the lowest score from the highest score.
6. **Variance**   The average squared deviation around the mean of a distribution.
7. **Standard Deviation**   The square root of the variance. The standard deviation is an estimate of the average deviation around the mean of a normal distribution.
8. **Population**   The entire collection of a set of objects, people, events, etc., of interest in a particular context.
9. **Sample**   A subset of observations selected from the population. Sampling should be done randomly to obtain a representative sample. In random sampling each case has an equal probability of being selected.

Exercises

The table on the following page lists the mathematics scores for 14 school districts in San Diego County.

1. Calculate the mean.
2. Calculate the median.
3. Is there a mode?
4. What is the range of scores?
5. Calculate the variance.
6. Calculate the standard deviation.

School	Mathematics Score
Coronado	72.6
Escondido	70.8
Fallbrook	71.7
Grossmont	70.6
Julian	68.7
Mountain Empire	64.1
Poway	74.3
Ramona	72.7
San Dieguito	75.0
Sweetwater	65.2
Vista	69.5
Carlsbad	69.2
Oceanside	66.8
San Marcos	69.0

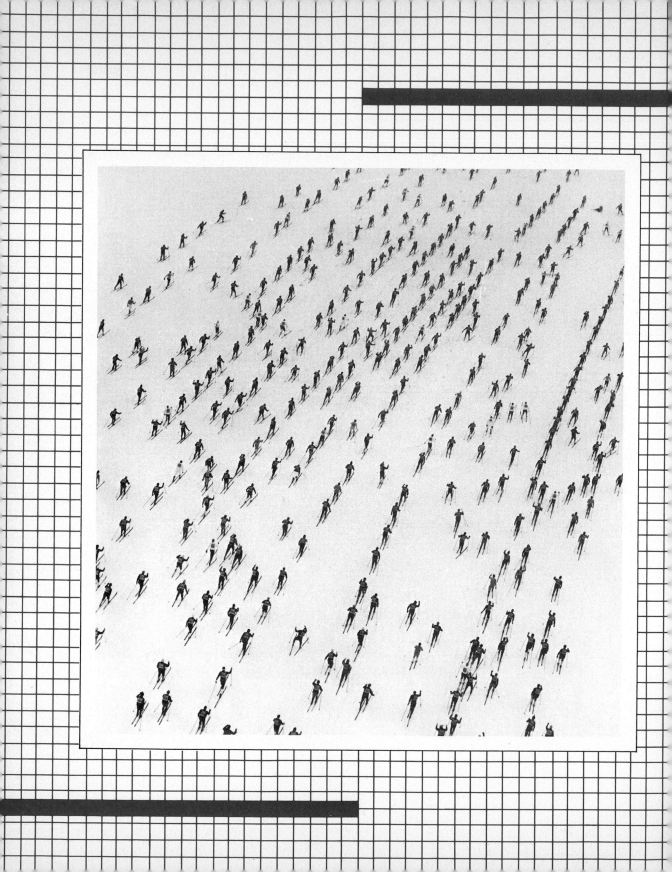

Finding Points Within Distributions

Comparisons abound in daily life. Perhaps one of the reasons number systems are common to all cultures is that they allow us to place observations into some context. Is the new car we are considering the most expensive, the least expensive, or somewhere in the middle of the distribution? A score of 27 on the physics exam is more satisfactory if 27 is in the top 10% of the scores than if it is in the bottom 10%.

One way to make comparisons among scores is to rank them. If we were to rank a group of skiers according to their performance on a particular day at Sun Valley, Idaho, we would assign the skier with the fastest time Rank 1, the skier with the second fastest time Rank 2, and so on. However, there is a problem with this system. It does not consider the number of people who are ranked. If you finish in 62nd place in a foot race, the meaning of your rank differs depending on the number of people participating in the event. If the race were only among 63 people, finishing in 62nd place may imply that you are a slow runner. However, some races have many, many participants. For instance, the Bay to Breakers in San Francisco has attracted as many as 50,000 entrants. If you finished in 62nd place, your performance would be considered very good.

Percentile Ranks

Percentile ranks replace simple ranks when we want to adjust for the number of scores in a group. A percentile rank expresses the percentage of scores that fall below a particular score (X_i). To calculate a percentile rank you need only follow these simple steps.

1. Determine how many cases are below the score of interest.
2. Determine how many cases are in the group.
3. Divide the number of cases below (Step 1) by the total number of cases in the group (Step 2).
4. Multiply the result of Step 3 by 100.

The formula is

$$P_r = \frac{B}{N} \times 100 = \text{percentile rank of } X_i, \qquad (4.1)$$

where B is the number of scores below X_i, and N is the total number of scores. This means that you form a ratio of the number of cases below the score and the number of scores. Because there will always be either the same or fewer cases in the numerator (top half) of the equation than there are in the denominator, this ratio will always be less than or equal to 1. To get rid of the decimal points we multiply by 100.

As an example, let us consider the runner who finishes 62nd out of 63 racers in a gym class. To obtain the percentile rank, we divide 1 (the number of people finishing behind the person of interest) by 63 (the number of scores in the group). This gives us 1/63, or .016. Then we multiply this result by 100 to obtain the percentile rank, which is 1.6. This rank tells us the runner is below the 2nd percentile.

In the Bay to Breakers example, if you had finished 62nd out of 50,000, then the number of people who were behind you would be 49,938. Dividing this by the number of entrants gives .9988. If we multiply by 100, we get the percentile rank, which is 99.88. This tells us that finishing 62nd in the Bay to Breakers foot race is exceptionally good. A runner finishing 62nd would be in the 99.88th percentile.

Calculations of Percentiles: Some Examples

Box 4.1 presents the calculation of percentile rank for the life expectancy of American men in comparision to men selected from other Westernized countries. Before proceeding we should point out that the meaning of this calculation depends on which countries are used in the comparison.

In this example, the calculation of the percentile rank is broken into five steps and uses the raw data in the table. In Step 1 we arrange the data points in descending order. Japan has the longest life expectancy at 72.9 years, Sweden

• BOX 4.1

Finding the Percentile Rank for Life Expectancy for Men in the U.S. in Comparison to That for Men in a Selected Sample of Countries

Formula: $P_r = \dfrac{B}{N} \times 100$

where P_r is the percentile rank, B is the number of cases below the case of interest, and N is the total number of cases.

COUNTRY	MALE LIFE EXPECTANCY
Canada	69.8
United States	69.3
Austria	68.5
Denmark	71.9
England/Wales	70.2
France	69.9
East Germany	68.9
West Germany	69.2
Ireland	69.0
Italy	69.9
Netherlands	72.2
Sweden	72.8
Switzerland	72.1
Israel	71.4
Japan	72.9
Australia	70.0
New Zealand	68.9

Source: U.S. Department of Health and Human Services. *Health in the United States,* 1980, p. 136.

STEPS

1. Arrange data in descending order—the highest score first, second highest score second, and so on.

COUNTRY	MALE LIFE EXPECTANCY
Japan	72.9
Sweden	72.8
Netherlands	72.2
Switzerland	72.1
Denmark	71.9
Israel	71.4
England/Wales	70.2
Australia	70.0
Italy	69.9
France	69.9
Canada	69.8
United States	69.3
West Germany	69.2
Ireland	69.0
East Germany	68.9
New Zealand	68.9
Austria	68.5

$N = 17$

$\bar{X} = 70.41$

$\sigma = 1.44$

2. Determine the number of cases below the score of interest. There are 5 countries in this sample with life expectancies lower than that in the U.S. (West Germany, Ireland, East Germany, New Zealand, and Austria).

3. Determine the number of cases in the sample. In this example there are 17.

4. Divide the number of scores below the score of interest (Step 2) by the total number of scores (Step 3).

$$\frac{5}{17} = .29$$

5. Multiply by 100.

$29 \times 100 = $ 29th percentile rank

is next with 72.8 years, and Austria has the shortest life expectancy at 68.5 years.

In Step 2 we determine the number of cases below the point of interest. In this example, the case of interest is the United States. Therefore, we count the number of cases below the United States. There are five (West Germany, Ireland, East Germany, New Zealand, and Austria). Each of these countries has a life expectancy shorter than the 69.3 years for the United States.

In Step 3 we determine the total number of cases. There are 17.

In Step 4 we divide the number of scores below the score of interest by the total number of scores:

$$\frac{5}{17} = .29.$$

Technically, the percentile rank is a percentage. Step 4 gives a proportion. Therefore, in Step 5 we transform this into a whole number by multiplying by 100, or

$$.29 \times 100 = 29.$$

Thus, the United States is in the 29th percentile.

The percentile rank depends absolutely on the cases used for comparison. In this example, we have calculated that the United States is in the 29th percentile for life expectancy within this group of countries. Had all of the countries in the world been included, the ranking of the United States might have been different.

Using this procedure, try to calculate the percentile rank for the Netherlands. The calculation is the same except that there are 14 countries below the Netherlands (as opposed to 5 below the United States). Thus, the percentile rank for the Netherlands is

$$\frac{14}{17} = .82 \times 100 = 82,$$

or the 82nd percentile. Now try England/Wales. You should get a percentile rank of 59.

Percentiles

Percentiles are the specific scores or points within a distribution. Percentiles divide the total frequency for a set of observations into hundredths.

Let us calculate the percentile and percentile rank for some of the data in Box 4.1. As an example, look at Israel. The life expectancy for Israeli men is 71.4 years. When calculating the percentile rank, we exclude the score and count those below it (in other words, Israel is not included in the count).

There are 11 countries in this sample with life expectancies of less than 71.4 years. To calculate the percentile rank we divide this number of countries by the total number of cases and multiply by 100, or

$$P_r = \frac{B}{N} \times 100 = \frac{11}{17} \times 100 = .65 \times 100$$
$$= 65.$$

Thus, we would say Israel is in the 65th percentile rank, or the 65th percentile in this example is 71.4 years.

Now let's take the example of Italy. This makes a good illustration because two countries have the same life expectancy (Italy and France). However, the calculation of percentile rank means that we look at the number of cases below the case of interest. In this example, there are seven cases less than 69.9 years (Canada, U.S., West Germany, Ireland, East Germany, New Zealand, and Austria). Thus, the percentile rank for Italy is $7/17 \times 100 = 41$. The 41st percentile corresponds with the point or score 69.9 years.

In summary, the percentile and the percentile rank are similar to one another. The percentile gives the point in a distribution below which a specified percentage of cases fall. The percentile is in raw score units. The percentile rank gives the proportion of cases below the percentile.

When reporting percentiles and percentile ranks, you must carefully specify the population that you are working with. Remember that a percentile rank is a measure of relative performance. When interpreting a percentile rank, you should always ask the question, "Relative to what?" Suppose for instance that you finished in the 17th percentile in a swimming race (or 5th in a heat of 6 competitors). Does this mean that you are a slow swimmer? Not necessarily. It may be that this was a heat in the Olympic games, and the participants were the fastest swimmers in the world. An Olympic swimmer competing against a random sample of all people in the world would probably finish in the 99.99th percentile.

Standardized Scores and Distributions

We can use the methods of the previous section to calculate the percentile and percentile rank for a score. However, in practice we often use other methods to locate a score within a distribution. These methods are less tedious and depend on our knowledge of commonly occurring distributions of scores. The next section requires an understanding of the mean and standard deviation. If these concepts seem fuzzy to you, you should go back and review the information in Chapter 3 before going on.

■ Z-Score

One of the problems with means and standard deviations is that their meanings are not clear. For example, if the mean of some set of scores is 57.6, it still does not convey all of the information we would like. There are other metrics designed for a more direct interpretation. The Z-score is a tranformation of data into standardized units that are easier to interpret. A Z-score is the difference between a score and the mean, divided by the standard deviation.

$$Z = \frac{X - \bar{X}}{S} \tag{4.2}$$

In other words, a Z-score is the deviation of a score X from the mean $\bar{X}$ in standard deviation units. If a score is equal to the mean, its Z-score is 0. For example, suppose the score and the mean are both 6; then $6 - 6 = 0$. Zero divided by anything is still 0. If the score is larger than the mean, the Z-score is positive; if the score is less than the mean, the Z-score is negative.

Let's try an example. Suppose that $X = 6$, the mean $\bar{X} = 3$, and the standard deviation* $S = 3$. Plugging these values into the formula we get

$$Z = \frac{6 - 3}{3} = \frac{3}{3}$$
$$= 1.$$

---

*Recall from Chapter 3 that the standard deviation is an approximation of the average deviation around the mean. Technically it is the square root of the average squared deviation around the mean and is defined by Equation 3.5.

Table 4.1

The calculation of mean, standard deviation, and Z-scores

NAME	TEST SCORE (X)	X^2
Carla	70	4,900
Fred	85	7,225
Monica	92	8,464
Andrew	65	4,225
James	83	6,889
Diane	98	9,604
Marcel	75	5,625
Sean	90	8,100
Jennie	60	3,600
Amanda	78	6,084
	$\Sigma X = 796$	$\Sigma X^2 = 64.716$

$$\overline{X} = \frac{\Sigma X}{N} = \frac{796}{10} = 79.60$$

$$S = \sqrt{\frac{\Sigma X^2 - \frac{(\Sigma X)^2}{N}}{N-1}} - \sqrt{\frac{64\,716 - \frac{(796)^2}{10}}{10-1}} = 12.267$$

$$\text{Monica's } Z\text{-score} = \frac{X - \overline{X}}{S} = \frac{92 - 79.6}{12.267} = 1.01$$

$$\text{Marcel's } Z\text{-score} = \frac{X - \overline{X}}{S} = \frac{75 - 79.6}{12.267} = -.37$$

$$\text{Jennie's } Z\text{-score} = \frac{X - \overline{X}}{S} = \frac{60 - 79.6}{12.267} = -1.60$$

Let's try another example. Suppose $X = 4$, $\overline{X} = 5.75$, and $S = 2.11$. What is the Z-score? It is $-.83$:

$$Z = \frac{4 - 5.75}{2.11}$$

$$= \frac{-1.75}{2.11}$$

$$= -.83.$$

This means that the score we observed (4) is .83 standard deviations below the average score, or that the score is below the mean, but its difference from the mean is slightly less than the average deviation. Some other examples are shown in Table 4.1.

■ Standard Normal Distribution

Now we consider the standard normal distribution because it is of central importance to statistics and to psychological testing. First you should participate in a short exercise. Take any coin and flip it 10 times. Now repeat this exercise of 10 coin flips 25 times. As you are doing this, record the number of heads you observe in each group of 10 flips. When you are done, make a frequency distribution showing how many times you observed 1 head in your 10 flips, 2 heads, 3 heads, and so on.

Your frequency distribution might look like the example shown in Fig. 4.1. The most frequently observed event occurs when you observe about equal numbers of 5 heads and 5 tails. As you go toward the 10 heads and 0 tails or 10 tails and 0 heads, events are observed with less frequency. For example, there were no occasions in which fewer than 2 heads were observed and only 1 occasion in which more than 8 heads were observed. This is very much what we would expect from the laws of probability. On the average we would expect half of the flips to show heads and half to show tails if heads and tails are equally probable events. Although it is still possible to observe a long string of heads or tails, it is improbable. In other words, we will sometimes observe that the coin comes up heads in 9 out of 10 flips. The likelihood that this will happen, however, is quite small.

Figure 4.2 shows the theoretical distribution of heads in an infinite number of flips of the coin. This figure might look a little like the distribution from

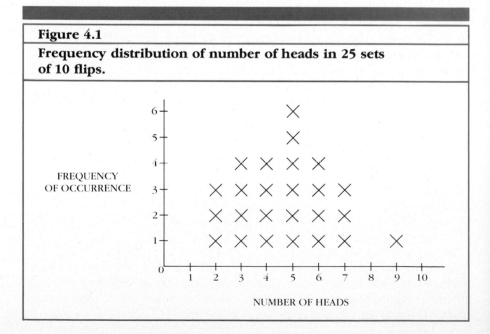

Figure 4.1

Frequency distribution of number of heads in 25 sets of 10 flips.

Figure 4.2

The theoretical distribution of the number of heads in an infinite number of coin flips.

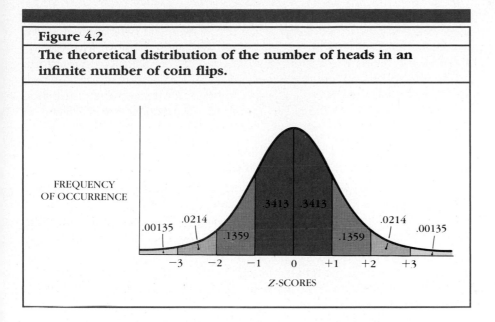

your coin-flipping exercise or the distribution shown in Fig. 4.1. Actually this is a normal distribution or what is known as a *symmetrical binomial probability distribution.*

On most occasions we refer to units on the x-axis of the normal distribution in Z-score units. Any variable transformed into Z-score units takes on special properties. First, Z-scores have a mean of 0 and a standard deviation of 1. If you think about this for a minute, you should be able to figure out why this is true. Recall that the sum of the deviations around the mean is always equal to 0. The numerator of the Z-score equation is just the deviation around the mean, while the denominator is a constant. Thus, the mean of Z-scores can be expressed as

$$\frac{\frac{1}{S} \Sigma(X - \overline{X})}{N}, \quad \text{or} \quad \frac{\Sigma Z}{N}. \tag{4.3}$$

Because $\Sigma(X - \overline{X})$ will always equal 0, the mean of Z-scores will always be 0. In Fig. 4.2 the standardized or Z-score units are marked on the x-axis. The numbers under the curve are proportions of cases in decimal form we would expect to observe in each area. Multiplying these proportions by 100 yields percentages. For example, we see that 34.13% or .3413 of the cases fall between the mean and 1 standard deviation above the mean. Do not forget that 50% of the cases fall below the mean. Putting these two bits of information together, we can conclude that if a score is 1 standard deviation above the mean, then it is at about the 84th percentile rank (50 + 34.13 = 84.13 to be exact). A score that is 1 standard deviation below the mean would be at about the 16th percentile rank (50 − 34.14 = 15.87). Thus, we can use what we have learned about means, standard deviations, Z-scores, and the normal curve to transform raw scores, which have very little meaning to us, into percentile scores, which are easier to interpret. These methods can be used whenever the distribution of scores is normal. Methods for nonnormal distributions are discussed in Chapter 12.

■ Percentiles and Z-Scores

Appendix 1 is a table that relates Z-scores to proportions. You can use this table to find the proportion of cases that fall above or below any Z-score. The Z-scores are listed in the column labeled Z. The second column gives the area from the mean to Z. (This area is a proportion and can be changed to a percentage by multiplying by 100.) The third column gives the area between negative infinity (the lowest possible point on the distribution) and the Z-score. The fourth column gives the area between the Z-score and positive infinity (the highest possible point).

Let's take a few examples. First consider a Z-score of 1.0 and find the Z-value in the table. (It occurs on the second page of the table.) The second column of the table shows that the area between the mean of the distribution and this Z-score is .3413. In other words, 34.13 percent of the distribution falls in this area.

Now let's consider the percentile rank for a Z-score of 1.0. This can be obtained from column three in Appendix 1. Column three shows the area between the bottom of the distribution and the observed Z-score. In the example of a Z-score of 1.0, the area is .8413. Since 1.0 is a positive Z-score, we know that at least .50 of the distribution lies below the Z-score because half of the Z-distribution lies below 0. Another way to have obtained the .8413 would have been to add .50 (the area below the 0, which is the mean of the Z-distribution) to .3413 (the area between the mean and the observed Z-score).

Now try to find the percentile rank of cases that fall below a Z-score of 1.10. If you are using the table correctly, you should obtain 86.43. Now try −.75. You should be aware that this is a negative Z-score, so the percentage of cases falling below the mean should be less than 50. You will notice that there are no negative values in Appendix 1. For a negative Z-score, there are several ways to obtain the appropriate area under the curve. The traditional way is to use column two in the table. This gives the area from the mean to the Z-score. In the example of a Z-score of −.75, we know that the area between the mean and the Z-score is .2734, or 27.34 percent of the distribution. In addition, we know that the mean is at the 50th percentile or that .50 of the cases fall below the mean. Thus, for a negative Z-score we can obtain the percentage of cases falling below the score by subtracting from .50 the tabled value listed in column two of Appendix 1. In this case it would be

$$.50 - .2734 = .2266.$$

Another way to solve the same problem would be to use column four in Appendix 1. The column shows the area from Z to the top of the distribution. However, we know that the distribution is symmetrical. Thus, if it is a negative Z-score, column four actually gives the area between a negative Z-score and the bottom of the distribution. The reason this is so is that the area above a positive Z-score will be exactly equal to the area below the same Z-score with a negative sign.

Consider a Z-score of $-.75$. What is the area below it? To obtain this value, we look at Appendix 1 and find the fourth column. The value is .2266. Multiplying this value by 100 gives the percentile rank. So 22.66 percent of the cases fall below the Z-score of $-.75$. It is very important that you learn to use Appendix 1. Practice with it until you are confident you understand how it works. Do not hesitate to ask your professor or teaching assistant if you are confused. This is a very important concept that you will need throughout the rest of the book.

We can turn the process around. Instead of using Z-scores to find the percentile ranks, we can use the percentile ranks, to find the corresponding Z-scores. To do this we need only look in Appendix 1 under percentiles and find the corresponding Z-score. For example, suppose we wish to find the Z-score associated with the 90th percentile. We look at column three of Appendix 1 and find that the value closest to the 90th percentile is 89.97, which lies adjacent to the Z-score of 1.28. This tells us that persons obtaining Z-scores greater than 1.28 are above the 90th percentile in the distribution.

■ An Example Close to Home

One of the difficulties in grading students is that performance is usually rated in terms of raw scores, such as the number of items a person gets correct on an examination. You are probably familiar with the experience of having a test returned to you with some number that makes little sense to you. For instance, the professor comes into class and hands you your test with a 72 on it. You must then wait patiently while he or she draws the distribution on the board and tries to put your 72 into some category that you understand, such as B+.

An alternative way of doing things would be to give you feedback about your performance as a Z-score. To do this, your professor would just subtract the average score (mean) from your score and divide by the standard deviation. If your Z-score was positive, you would immediately know that your score was above average; if it were negative, you would know your performance was below average.

Table 4.2		
Z-score cutoffs for a grading system		
GRADE	PERCENTILES	Z-SCORE CUTOFF
A	85–100	1.04
B	60–84	.25
C	20–59	−.84
D	6–19	−1.56
F	0–5	Less than −1.56

Suppose that your professor tells you in advance that you will be graded on a curve system, according to the following rigid criteria. If you are in the top 15% of the class, you will get an "A" (85th percentile or above); between the 60th and the 84th percentile, you will get a "B"; between the 20th and the 59th percentile, you will get a "C"; between the 6th and the 19th percentile, you will get a "D"; and in the 5th percentile or below, you will not pass. Using Appendix 1, you should be able to find the Z-scores associated with each of these cutoff points. Try it on your own, and then consult Table 4.2 to see if you are correct.

Looking at Table 4.2 you should be able to determine what your grade would be in this class on the basis of your Z-score. If your Z-score is 1.04 or greater, you would receive an "A". If it were greater than .25 but less than 1.04, you would get a "B", and so on. This system assumes that the scores are normally distributed.

Now let us try an example that puts a few of the concepts together. Suppose that you get a 60 on a social psychology examination. You learned in class that the mean for the test was 55.70 and that the standard deviation was 6.08. If your professor uses the same grading system that was just described, what would your grade be?

To solve this problem we must first find your Z-score. The formula for a Z-score is given in Equation 4.2:

$$Z = \frac{X - \bar{X}}{S}. \tag{4.2}$$

So your Z-score would be

$$Z = \frac{60 - 55.70}{6.08} = \frac{4.30}{6.08} = .707.$$

Looking at Table 4.2, we find that .707 is greater than .25 (the cutoff for a "B") but less than 1.04 (the cutoff for an "A"). Now let us find your exact standing in the class. To do this we need to look again at Appendix 1. Because the table gives Z-scores only to the second decimal, we round .707 to .71. Looking at the appendix, we find that 76.11% of the cases fall below a Z-score of .71.

This means that you would be in approximately the 76th percentile or that you would have performed better on this examination than approximately 76 out of every 100 students.

■ McCall's *T*

There are a variety of other systems by which we can transform raw scores to give them more intuitive meaning. One system was established in 1939 by W. A. McCall, who originally intended to develop a system to derive equal units on mental quantities. He suggested that a random sample of 12-year-olds be tested and that the distribution of their scores be obtained. Then percentile equivalents were to be assigned to each raw score, showing the percentile rank in the group for the persons obtaining that raw score. After this had been accomplished the mean of the distribution would be set at 50, to correspond with the 50th percentile. In McCall's system, that standard deviation was set at 10.

In effect, what McCall generated was a system that is exactly the same as standard scores (Z-scores) except the mean in McCall's system is 50 rather than 0 and the standard deviation is 10 rather than 1. Indeed, a Z-score can be transformed to a T-score by applying the linear transformation

$$T = 10Z + 50. \tag{4.4}$$

In words, you can get from a Z-score to McCall's T by multiplying the Z-score by 10 and adding 50. It should be noted that McCall did not originally intend to create an alternative to the Z-score. He wanted to obtain one set of scores which could then be applied in other situations without standardizing the entire set of numbers.

As many people have pointed out (for example, Angoff, 1971), there is nothing magical about the mean of 50 and the standard deviation of 10. It is a simple matter to create systems such as standard scores with any mean and standard deviation you like. If you want to say that you got a score 1000 points higher than a person who was 1 standard deviation below you, you could devise a system with a mean of 100,000 and a standard deviation of 1000. If you had calculated Z-scores for this distribution, you would obtain this with the transformation

$$NS \text{ (for New Score)} = 1000Z + 100,000.$$

In fact, you can create any system desired. To do so, just multiply the Z-score by whatever you would like the standard deviation of your distribution to be and then add the number you would like the mean of your new distribution to be.

An example of a test that was developed using standardized scores is the Scholastic Aptitude Test (SAT). When this test was created in 1941, the developers decided to make the mean score 500 and the standard deviation 100. Thus, they multiplied the Z-scores for those who took the test by 100 and added 500. Since the test has been in use, the basic scoring system has not been changed.

Quartiles and Deciles

The terms **quartiles** and **deciles** are frequently used when tests and test results are discussed. The two terms refer to divisions of the percentile scale into groups. The quartile system divides the percentile scale into four groups, and the decile system divides the scale into ten groups.

The first quartile is the 25th percentile, the second quartile is the median or the 50th percentile, and the third quartile is the 75th percentile. These are abbreviated Q_1, Q_2, and Q_3, respectively. One-quarter of the cases fall below Q_1, one-half fall below Q_2, and three-quarters fall below Q_3. The **interquartile range** is the interval of scores bounded by the 25th and 75th percentiles. In other words, the interquartile range is bounded by the range of scores that represents the middle 50% of the distribution.

Deciles are similar to quartiles except that they use points that mark 10% rather than 25% intervals. Thus, the top decile, or D_9, is the point at which 90% of the cases fall below. The next decile D_8 marks the 80th percentile, and so forth.

Summary

This chapter reviews methods for finding points within distributions of scores. One of the most commonly used methods is the percentile rank, which is the proportion of the total number of scores that fall below a particular score. The percentile point, or percentile, is the point in the distribution that marks the cutoff for a specific percentile rank. For example, the median of a distribution marks the 50th percentile. Suppose that the median of a distribution of scores is 86. The percentile rank for a score of 86 would be 50. The percentile point for the 50th percentile would be 86.

Standardized scores, or Z-scores, are used to place points in distributions into context. The specific context is the standard normal distribution. Using standardized tables we can convert Z-scores into percentile ranks. For example, a Z-score of 1.0 is associated with a percentile rank of 84.13. A Z-score is equal to the deviation of a score from the mean of its distribution divided by the standard deviation.

For some applications we use special transformations of Z-scores. For any distribution, the mean of Z-scores is zero and the standard deviation is 1. One common transformation of Z-scores is McCall's T. In McCall's T-scores the mean is 50 and the standard deviation is 10. Thus, a T-score of 60 is 1 standard deviation above the mean.

In other special applications we use predefined percentile points. For example, deciles are points that divide the frequency distribution into 10 equal intervals. Quartiles are points that divide the frequency distribution into 4 equal intervals.

Key Terms

1. **Percentile Rank**   An expression of the proportion of scores that fall below a particular score.
2. **Percentile Points**   Points in distributions associated with particular percentile ranks.
3. **Z-Score**   The difference between a score and the mean, divided by the standard deviation. The Z-score is a deviation score expressed in standard deviation units.
4. **Standard Normal Distribution**   A probability distribution obtained from an infinite sequence of binomial (two-choice) chance events. The standard normal distribution is used for reference in statistics because its properties are well understood.
5. **Quartiles**   Points that divide the frequency distribution into equal fourths. The first quartile is the 25th percentile, the second quartile is the median, the third quartile is the 75th percentile.
6. **Interquartile Range**   The interval bounded by the 25th and 75th percentiles.
7. **Deciles**   Points that divide the frequency distribution into 10 equal intervals. For example, the top decile, D_9, is the point at which 90% of the cases fall below.

Exercises

The table on the following page lists the mathematics scores for 14 school districts in San Diego County.

1. In Chapter 3, you calculated the mean and the standard deviation for these scores. Using the mean and standard deviation (if you did not save your results, look them up in the back of the book) find the Z-score for Fallbrook. Now, find the Z-score for Carlsbad and for San Dieguito.
2. What is the percentile rank for Vista?
3. Which schools are in the upper quartile?

School	Mathematics Score
Coronado	72.6
Escondido	70.8
Fallbrook	71.7
Grossmont	70.6
Julian	68.7
Mountain Empire	64.1
Poway	74.3
Ramona	72.7
San Dieguito	75.0
Sweetwater	65.2
Vista	69.5
Carlsbad	69.2
Oceanside	66.8
San Marcos	69.0

4. Which schools are in the bottom decile?

5. What is the interquartile range for this distribution of scores?

Inferential Statistics

II

Many problems in statistics require us to estimate or make educated guesses on the basis of numerical information. For example, we might want to estimate the average IQ of American doctors. Since we are not capable of administering an IQ test to every doctor, we administer the test to a small number, calculate the average, and estimate what the score may have been for the entire group.

These estimations are, of course, subject to error. Inferential statistics are used to make educated guesses that take the chances of making errors into consideration. In order to apply these methods, principles of probability theory must be used. A summary of probability theory is presented in Chapter 5. Chapter 6 is a theoretical overview of statistical inference methods. Chapters 7, 8, and 9 give specific methods for making statistical inferences. The methods covered in Chapter 7 are used to compare two groups of scores, including the Z-test and the t-test. The method called the analysis of variance, one typically used to compare three or more groups, is presented in Chapter 8. More complex models of the analysis of variance are presented in Chapter 9.

5

Introduction to Probability

Gambles in Everyday Life

Life is a series of gambles. Although we are sometimes unaware of it, nearly all decisions require an assessment of probabilities. A probability is a quantitative statement of the likelihood that an event will occur. A probability of 0 means the event is certain not to occur; a probability of 1.0 means the event will occur with certainty. Some people feel there are no events in the world with probabilities of 0 or 1.0. However, there are some good bets. For instance, we can say with almost absolute certainty (probability = 1.0) that the sun will rise tomorrow. There are also events for which we can assume the probability is very close to 0, for instance, the probability that the gas and electric company will roll back prices to their 1960 levels. We cannot say with absolute certainty that these events will occur, but we might operate as though we knew them as fact.

For most decisions, we must use some estimate of probability that is between 0 and 1.0. The most interesting events in life are those where there is some element of uncertainty. Football fans estimate the probability that their team will make the Superbowl. College students assess their chances of getting good grades in different classes or of being accepted to graduate school.

Many decisions are bets against uncertainty. For example, an insurance policy provides assurance that you will receive compensation for a very unlikely event. To make money, the insurance company has estimated the probabilities that they will have to make a payment, and they have assessed the rates so that they can make a profit after making payments to a certain number of policyholders.

The preceding chapters have discussed descriptive statistics. These statistics are used primarily to describe distributions of scores. The rest of the book is devoted to *inferential* statistics. Inferential statistics are used to make inferences or general statements about a population based on a sample from that population. For example, if you wanted to find the average height of American men, you might get a measuring tape and measure each and every American man. However, this could be tedious and very costly. Another approach would be to take a random sample of American men and measure them. The number of men in this random sample would be considerably smaller than the total population of men. The major concern is whether the sample mean is equivalent to the population mean. Of course, we would not know this unless both figures were available. Inferential statistics are used to estimate the degree of correspondence between these two. We will never be able to say with certainty that the two means are exactly the same, but we will be able to state the probability that they differ significantly.

All of the statistical tests covered in the remainder of this book allow us to make probability statements. Consider this statement: a population mean falls within some defined interval around a sample mean. To help us understand this and other statements, let's take time out for a brief review of probability theory.

Basic Terms

Probability theory deals with defined laws of chance. In a random series of events, there are known regularities in observed proportions. Originally, probability theory was used to describe regularities in the observed proportions in various games of gambling. For example, early probability theorists considered the chances of winning in games of roulette. Soon scientists began to realize that understanding the laws of chance or probability theory could help them gain insights into nature and the universe.

To understand probability theory, it will be necessary for you to master some basic terminology. Following is a glossary of terms that will help you understand the discussion.

Random Experiment: In everyday life and in science, we are likely to "try things out." For example, a drug may be injected into chickens and observations made to determine how the drug affected behavior. In other experiments, comparisons may be made among different methods of learning, or we may wish to determine whether people are more aggressive after they have watched television. In science, we strive to construct experiments so that the results are replicable. This means that different scientists using the same method should obtain the same results. When the same condition produces the same result, we say that the experiment is *deterministic*. In other words, the experimental conditions determine the result. However, all experiments do not produce exactly the same result. In the behavioral and social sciences, it would actually be rare to get exactly the same data when experiments are repeated. The extent to which the results of identical experiments differ is considered to be random. A random outcome is one determined completely by chance.

In a random experiment, the results are determined completely by chance. Thus, in repeated random experiments, the results differ and the distribution of results reflects a normal probability distribution. For example, flipping a coin is a random experiment. Each coin flip is an experiment for which the probability of various results is random.

Set: A set is a collection of things or objects that are clearly defined by some rule. For example, all sophomore women at Indiana University define a set. Another set might be all patients with diabetes living in California. Within sets, there may also be subsets. For example, women psychology students who are sophomores at Indiana University are a subset of the set of sophomore women at Indiana University. Diabetic patients living in San Diego are a subset of the set of diabetic patients in California.

Element: An element is any member within a set. For example, Susan Schwartz is a sophomore psychology major at Indiana University. Thus, she is an element in the set of sophomore women at Indiana University and the subset sophomore women psychology students at Indiana University.

Empty Set: An empty set is a set with no elements. For example, consider the set of female players on the San Diego Chargers professional football team. Since there are no female players, the set is empty.

Union: For two sets A and B, the union is all elements that are in A, in B, or in both A and B. (See Fig. 5.1.) For example, consider the group of psychology and political science majors at a state university. Suppose that some students have a joint major of psychology and political science. The union includes all of those majoring in psychology, all of those majoring in political science, and those majoring in both psychology and political science. Union is symbolized as $A \cup B$.

Intersection: For sets A and B, the intersection is the subset of all elements that are in both A and B, but not in A alone or in B alone. The intersection of A and B is symbolized as $A \cap B$. (See Fig. 5.2.)

Mutually Exclusive Events: Two events are mutually exclusive if they share no common elements. In a sample of college students, each individual is either male or female. Since no individual is male and female, we say that the genders are mutually exclusive. Similarly, each player in the National Football League plays for only one team. Thus, team membership is mutually exclusive. (See Fig. 5.3.)

Complement: Some sets contain subsets. If there is a subset, the complement is made up of all other elements in the set outside of the subset. For example, if P is a subset of set U, we can define the complement of P as $\bar{P}$. Thus, $\bar{P}$ contains all elements of the set that are not P. For example, suppose that the set is deaths due to all diseases in a certain year. The subset P might be deaths due

Figure 5.1

Union.
The total shade area in both circles is
the union of P and Q, or $P \cup Q$.

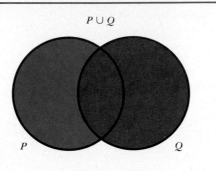

$P \cup Q$

P Q

Figure 5.2

Intersection.
The shaded area is the intersection of P
and Q, or $P \cap Q$.

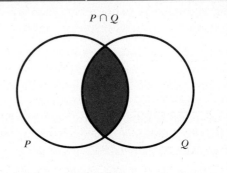

$P \cap Q$

P Q

Figure 5.3

Mutually exclusive events.
The set of elements in P and Q is empty
(symbolized as 0). Thus, $P \cap Q = 0$.

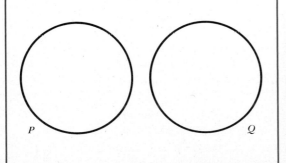

P Q

Figure 5.4

Complementary events.
In this universe, every element is either
P (open circle) or not P ($\bar{P}$, indicated by
shaded area).

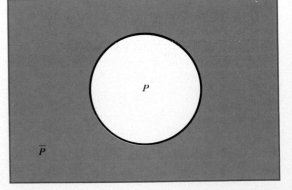

P

$\bar{P}$

to heart disease. We might now want to lump all other deaths together as not heart disease. These would be $\bar{P}$, which is the complement of P. (See Fig. 5.4.)

Probability is the study of odds or chances. If you flip a coin, there are two potential outcomes: heads (H) or tails (T). On a single flip of the coin, the probability of each outcome is 1/2 or .5. Simple probabilities are calculated by dividing the specific outcome by the number of possible outcomes. For example, the probability of getting a head in a single coin flip is

$\dfrac{1}{2}$ ⟵ Heads
⟵ All possible outcomes (Heads + Tails)

The situation becomes more complex when you flip the coin twice. For two coin tosses, there are four potential outcomes: HH, HT, TH, TT. The possible results of three coin tosses are shown in Table 5.1.

Look at Table 5.1. What is the probability of getting at least one head in three tosses? The answer is 7/8. The table shows there are eight potential outcomes, and heads occur in seven of the eight possibilities. Now what is the probability of getting a head on the first toss and a tail on either the second or the third toss? The answer is 3/8. There are three possible ways to achieve this outcome: HHT, HTH, HTT.

What is the probability of getting at least two heads? The probability is 4/8 (HHH, HHT, HTH, THH). Usually we express probabilities as numbers between 0 and 1. So the fractions 1/4, 3/8, etc., can be translated into decimals

Table 5.1

OUTCOME OF FIRST TOSS	OUTCOME OF SECOND TOSS	OUTCOME OF THIRD TOSS
Head	Head	Head
Head	Head	Tail
Head	Tail	Head
Head	Tail	Tail
Tail	Head	Head
Tail	Head	Tail
Tail	Tail	Head
Tail	Tail	Tail

by dividing the numerator by the denominator. For an event with a probability of 3/8, we would say that its probability is .375; or the chances of getting at least one head in three tosses of a coin are 7/8 or .875.

The number of possible outcomes is found by taking the number of potential outcomes on a single toss and raising it to the power of the number of tosses. For example, the number of potential outcomes for a single coin toss is

$$2^1 = 2.$$

The number of potential outcomes for three tosses of a coin is

$$2^3 = 8.$$

For five coin tosses it is

$$2^5 = 32,$$

and for ten tosses it is

$$2^{10} = 1024.$$

Let's take the example of 5 independent and consecutive coin tosses. The number of potential outcomes is 32 ($2^5 = 32$). Thus, the probability of any specific result would be 1/32, or .03. However, each of these outcomes represents a specific order of events. For instance, the combination TTHHH and the combination HHHTT are both considered independent events. Yet in each case the outcome is three heads and two tails. Most often we are interested only in the outcomes over many trials. Thus, different ways of obtaining the same outcome might be lumped together. For example, the outcome of three heads and two tails could be obtained in ten different ways. These are shown in Table 5.2. If order is not important, then 10 of the 32 possible outcomes

Table 5.2					
Different ways to obtain three heads and two tails in five flips of an unbiased coin					
TRIAL	FLIP 1	FLIP 2	FLIP 3	FLIP 4	FLIP 5
1	H	H	H	T	T
2	H	H	T	H	T
3	H	H	T	T	H
4	H	T	H	H	T
5	H	T	T	H	H
6	H	T	H	T	H
7	T	H	H	T	H
8	T	H	T	H	H
9	T	H	H	H	T
10	T	T	H	H	H

Figure 5.5

Pascal's triangle.

Each row begins and ends with a 1. Values between the 1's are obtained by adding the two adjacent values in the row above. For example, the 2 in the third row was obtained by adding the two adjacent values in the row above (1 + 1). There are two 10's in the sixth row. Each was obtained by adding the two adjacent values from the row above (4 + 6 for the first 10 and 6 + 4 for the second 10). Within each circled area (part b) you can see how adjacent numbers in a row were added to obtain the number in the new row.

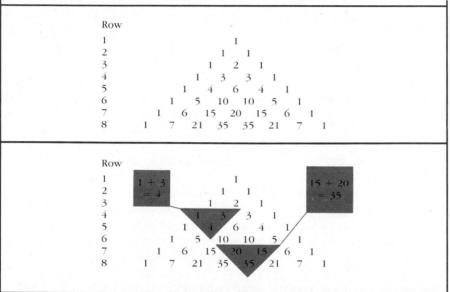

from 5 consecutive flips of a coin can be grouped together as three heads and two tails. Similarly, there are ten possible combinations of three tails and two heads.

Several centuries ago Pascal worked out a method for displaying all of the possible outcomes for dichotomous events. A dichotomous event is one that has two potential outcomes. Pascal summarized his findings in a simple table that has come to be known as Pascal's Triangle. Figure 5.5 is an example of Pascal's Triangle for 8 trials.

Notice that the number 1 is represented on the diagonal on each side of the triangle. To obtain all of the other entries in the figure, sum any two adjacent numbers in a row to get the number halfway between the numbers on the next line. For example, the numbers in the second row are 1 and 1. Summing these, we get 2, which is located in the next row down between the two 1's in the second row. Let's try a more difficult example. Find the numbers 4 and 6 in the fifth row. Add them to get 10. Now look at the number halfway between 4 and 6 in the 6th row of the figure. It is 10. Experiment with the figure to be sure you understand these arrangements.

Using Pascal's Triangle is one method for finding probabilities of an event. Summing across any row will give the number of potential outcomes for a dichotomous event for a defined number of independent trials. For example, the fifth row sums to the number of potential outcomes for 4 independent dichotomous events:

$$1 + 4 + 6 + 4 + 1 = 16 = 2^4.$$

The probability of any one outcome is 1 divided by the number of potential outcomes. Pascal's Triangle can be used to obtain the probability of $1/16 = .06$ for this example. However, there are more systematic ways, which we consider in the next section. Before doing so, it will be necessary to point out the importance of considering events as independent of one another. In a series of coin flips we usually assume that the events are independent. This means that the outcome on any trial has no influence on the outcome of any other trial. This happens if the coin is unbiased. Indeed, if we found in repeated trials that we obtained heads in 75% of the coin flips, we might conclude that the coin was biased.

Imagine that you are betting on the outcomes in a series of coin flips. You know that the coin is unbiased, but it has come up heads three consecutive times. Further, you realize that this is improbable (the chances of three consecutive heads are 1 in 8), so you bet on heads the next trial. Many people feel that if a coin has come up heads several times consecutively, it should come up tails the next time. This is what is known as *gamblers' fallacy*. If the coin is unbiased, the chances of it coming up heads are exactly the same in each trial, independent of what has happened in preceding trials. Even though the coin may have come up heads three consecutive times, the chances of it coming up heads in the next trial are also $1/2$. Gamblers' fallacy is the expectation that history will affect the next trial. In fact, in independent trials each flip of the coin is a new ball game.

There are many situations in which independent trials are not truly independent. For example, consider the chances that the University of Southern California will win its next football game. Even though each football game has a dichotomous outcome (win or loss), these events are not independent. The probability of winning the next game may be related to the abilities of the team, which also influenced preceding games. Consider the chances of a team winning the sixth game of the season after winning the first five. If the outcome of the game were a chance event, the probability of winning the sixth game would be .5, just as it would have been for all preceding games. However, the outcome of a football game is probably not an independent event. Thus, sterling performance in previous games suggests a greater likelihood that the team will win the next game.

Additive and Multiplicative Rules

Many problems in statistics and probability require us to combine two independent probability estimates. Many times it is difficult to determine how to combine independent probabilities to make a joint probability statement.

For example, suppose that you were randomly selecting cards from a standard 52-card deck. You draw a single card and you want to know the probability that the card is an 8 or a king. According to the game you are playing, you will win if you can get either, but you need to know the probabilities before you offer a bet.

Sometimes it is difficult to determine how to figure these probabilities. However, there are some simple rules. First, there is the additive rule, which is sometimes called the "or" rule. When we are trying to determine the probability of one event *or* the other, we simply add the two probabilities together. For example, if we want to determine the chances of obtaining either an 8 or a king in a single draw from a 52-card deck, we *add* the two independent probabilities. The probability of drawing an 8 is 1/13, while the probability of drawing a king is also 1/13. Therefore, the probability of drawing either an 8 or a king is

$$\frac{1}{13} + \frac{1}{13} = \frac{2}{13}$$

$$= .15.$$

Now let's take an example using Z-scores. Suppose that we are going to randomly select a person from the general population and measure her IQ. We want to know the probability that she will be either one standard deviation above or two standard deviations below the mean in IQ. First we determine the probability of being one standard deviation above the mean. We look in Appendix 1 for the Z-score 1.0. The table shows us that the probability of being one standard deviation above the mean is .16 (see Column 4; .1587 rounds to .16; if you have trouble finding this you should go back and review the section on Z-scores in Chapter 4). Now we want to find the probability of being two standard deviations below the mean. According to Appendix 1 that probability is .02. Since we want to know the probability of selecting someone who is one standard deviation above *or* two standard deviations below we use the additive rule and add the two independent probabilities.

$$.16 + .02 = .18$$

---

*This includes the probability of getting both a six and a head.

Thus, the chances are 18 in 100 that we would select someone one standard deviation above or two standard deviations below the mean in IQ. In summary, the additive rule expresses the probability of the union.

Now let's consider the intersection. Suppose we want to know the probability that an event is in both A and B. Consider the chances of obtaining both a head *and* a six in the roll of a die and the flip of a coin. When calculating the probability of a joint occurrence we use the multiplication rule. The multiplication rule instructs us to multiply the independent probabilities of the two events. In this case we multiply the probability of obtaining a head in the coin flip (1/2) by the probability of getting a six in the roll of a die (1/6).

$$\frac{1}{2} \times \frac{1}{6} = \frac{1}{12}$$
$$= .08$$

The chances are 8 in 100.

Let's take another example. Suppose that we randomly select two children from the general population of children. We want to know the probability of both children having an IQ greater than or equal to two standard deviations above the mean. To make this calculation we first check Appendix 1 to find the probability of having a score greater than or equal to two standard deviations above the mean (or a Z-score greater than or equal to 2.0). That probability is approximately .02. Our problem can be stated as follows: what are the chances of selecting a first child with an IQ greater than or equal to two standard deviations above the mean *and* sampling a second child with an equally high IQ? The key for the use of the multiplicative rule is the word *and*. Because the word *and* is in the statement we multiply the two probabilities. The probability of this event is

$.02 \times .02 = .0004.$

In other words, the probability is very small that we would randomly select two children with IQ's greater than or equal to two standard deviations above the mean.

Permutations and Combinations

There are some situations in which both the additive and the multiplicative rules apply. Specifically, both rules apply when we want to determine the probability that there will be a certain number of occurrences in a given number of trials. These situations require special methods of counting known as permutations and combinations. In the preceding section, you learned how to determine simple probabilities. However, it can be tedious to list all the potential outcomes and then determine the proportion that fall within the some defined category. This is particularly true if there are many possible outcomes. Fortunately there are precise mathematical formulas for calculating these probabilities. In the next section we will explore permutation and combination methods.

Permutations and combinations are methods for counting. The difference between permutations and combinations is in the importance of order. Permutations require predicting the precise order of an outcome; combinations do not consider order at all. Consider a simple card game that involves only the aces from a deck. There are four aces: spades, clubs, hearts, and diamonds. Now suppose that you shuffle these four cards and then draw two. What is the probability of drawing a heart followed by a diamond? To determine this we must list all possible outcomes. (We will refer to spades as S, clubs as C, hearts as H, and diamonds as D.) There are 12 possible outcomes:

SC, SH, SD, CS, CH, CD, HC, HS, HD, DC, DS, DH.

Note that the outcomes do not include CC, SS, HH, or DD. This is because once a card is drawn it is taken out of circulation.

There is only one way that the specified two cards can be drawn. As you can see the chances are 1/12. Look at this another way. The chances of drawing a heart on the first draw are 1 in 4. Once the heart has been drawn, consider the chances of drawing a diamond next. Since there are only 3 cards remaining, the probability is 1/3. Thus, the probability is

$$\frac{1}{4} \times \frac{1}{3} = \frac{1}{12}.$$

Drawing the cards in a specific order is a permutation.

■ Factorials

Consider the same problem but ignore having to draw the cards in a specific order. We assume that drawing a heart and a diamond is sufficient, and it does not matter which we draw first. That is, drawing a heart and then a diamond is equivalent to drawing a diamond and then a heart, or HD = DH. When we do this, the number of possible combinations reduces to 6:

SC, SH, SD, CH, CD, HD.

Thus, the probability of drawing a heart and a diamond is 1/6 or .167. This approach, which does not consider order, is called **combinations.** Calculating permutations and combinations the way we have is very tedious. Therefore, mathematicians developed formulas to find permutations and combinations.

To understand these formulas you must review the concept of a *factorial*. The factorial for a number is the product of the integers from 1 to the number. The factorial is signified by an exclamation point. For example, the factorial of 6 is expressed as

$$6! = 6 \times 5 \times 4 \times 3 \times 2 \times 1$$
$$= 720.$$

Again, to find the factorial we multiply the number times each successive integer below it. This formula works for all positive integers. Zero is a special case. The factorial of zero is 1.

It is worth noting that the factorials for numbers grow rapidly as the numbers increase. For example,

$$1! = 1,$$
$$2! = 2,$$
$$3! = 6,$$

$$\vdots \qquad \vdots$$

$$8! = 40,320,$$

$$\vdots \qquad \vdots$$

$$15! = 1,307,674,368,000.$$

In other words, as the magnitude of an integer increases, the magnitude of its factorial increases dramatically. The factorial of 100, for example, is a number that is so large it is beyond our comprehension.

■ Permutations

Now that you have learned about factorials, you can use formulas to calculate permutations and combinations. The formula for permutations is

$$_NP_M = \frac{N!}{(N - M)!} \tag{5.1}$$

On the left side of the equation is the sign for permutations. The subscripts to the left and right of the P tell the number of objects and how many of them are taken at a time. N is the number of objects, and M is the number of these objects taken one at a time. For example,

$$_8P_2$$

means we are taking 8 objects, 2 at a time. Consider the example of the number of different possible hands you could obtain if you drew two cards from our four-card deck. Recall that the four-card deck included the aces of spades, clubs, hearts, and diamonds. We already know from the earlier discussion that there are 12 permutations. Let's see if we can get the same result using the permutations formula. Since there are four cards, $N = 4$. In other words, N, the number of things, equals four. The other bit of information we need is M, which is the number of things taken at a time. Since we are drawing hands of two cards, $M = 2$. Now we can put this information into the formula:

$$_NP_M = \frac{N!}{(N - M)!} = {}_4P_2 = \frac{4!}{(4 - 2)!}.$$

We can simplify the denominator by subtracting $4 - 2$. This reduces to

$$_4P_2 = \frac{4!}{2!}.$$

Now we are ready to multiply out the factorials:

$$_4P_2 = \frac{4 \times 3 \times 2 \times 1}{2 \times 1} = \frac{24}{2} = 12.$$

To summarize, we have used the permutations formula to calculate the number of different specific ordered pairs of cards that can be drawn from a deck of four cards.

Actually, the process can be simplified because algebra rules make it possible to cancel some of the numbers in permutations formulas. Look at the preceding formula:

$$\frac{4 \times 3 \times 2 \times 1}{2 \times 1}$$

Notice that 2×1 is in both the top and bottom of the formula. Using the rules of algebra, these identical quantities can be circled, and then we can cancel these two out.

$$\frac{4 \times 3 \times \cancel{2} \times \cancel{1}}{\cancel{2} \times \cancel{1}} = 4 \times 3 = 12$$

Thus, the complex procedure for calculating permutations is somewhat simplified.

Box 5.1 shows the step-by-step calculation of permutations for the example of 5 things taken 2 at a time. This might represent scheduling games for a basketball league that has 5 teams. Five teams are the 5 "things." Suppose that you are in charge of the scheduling of games, and you want each team to play each other team. In addition, you consider home court advantage to be important, so each team plays each other team both at home and away. Thus, scheduling Team A vs. Team B is a separate event from scheduling Team B vs. Team A. As the calculations in Box 5.1 show, you would need 20 games to accomplish this objective.

● BOX 5.1

Step-by-Step Calculation of Permutations

5 THINGS TAKEN 2 AT A TIME

Permutations:

$$_NP_M = \frac{N!}{(N-M)!}$$

$$_5P_2 = \frac{5!}{(5-2)!}$$

STEPS

1. Reduce the denominator by subtracting M from N.

$$5 - 2 = 3, \quad \text{so } _5P_2 = \frac{5!}{3!}$$

2. Write out the factorials for the numerator and the denominator.

$$\frac{5!}{3!} = \frac{5 \times 4 \times 3 \times 2 \times 1}{3 \times 2 \times 1}$$

3. Cancel.

$$\frac{5!}{3!} = \frac{5 \times 4 \times \cancel{3} \times \cancel{2} \times \cancel{1}}{\cancel{3} \times \cancel{2} \times \cancel{1}}$$

4. Multiply the remaining numbers.

$$5 \times 4 = 20$$

■ Combinations

As noted earlier, combinations are similar to permutations except that the order is not important. For our basketball team example, imagine that it was unimportant where the two teams met. (In the permutation example, we consider Team A meeting Team B on Team A's court a different event than meeting on Team B's court.) In combinations, we consider only that the teams met. The formula for combinations is

$$_NC_M = \frac{N!}{M!\,(N-M)!}. \tag{5.2}$$

In the left side of the formula we are finding the number of objects (N) that can be selected from another group of objects (M) without considering order.

To illustrate combinations, let's return to the example of drawing 2 cards from a deck of 4 cards. This time the order in which we draw the cards is not important. For example, drawing a heart and a club is considered the same as drawing a club and a heart. When we plug the numbers into the formula, we get

$$_4C_2 = \frac{4!}{2!\,(4-2)!}$$

$$= \frac{4 \times 3 \times \cancel{2} \times \cancel{1}}{(2 \times 1)\,(\cancel{2} \times \cancel{1})}$$

$$= \frac{12}{2}$$

$$= 6.$$

In words, there are 6 different ways to draw 2 cards from 4 cards when order is unimportant.

Box 5.2 works through more complex examples of permutations and combinations. The example considers the chances of winning the perfecta at the race track. Some race tracks have a featured race in which the bettor must name the first 3 horses to complete the race. Let us suppose that there is a small field of only 8 horses. Further suppose that the bettor is unwilling to trust any information in the racing form and that he or she assumes that all horses have an equal chance of winning. If indeed all horses have an equal chance of winning, and our bettor makes a random bet on the perfecta, what are the chances of his or her winning? The answer is less than 3 in 1000. Stated another way, for each 1000 bets a gambler makes, he or she would expect to win 3 by chance. The second portion of Box 5.2 considers the same problem but for combinations.

In this case we create the hypothetical situation in which the bettor must pick 3 horses of 8 to come in first, second, and third. However, the order in which these horses finish is irrelevant. In other words, if horse 1 finished first, horse 2 finished second, and horse 3 finished third, this would be considered equivalent to horse 1 coming in third, horse 2 coming in second, and horse 3 coming in first. As the box shows, when we are not concerned about the exact order of the three horses finishing, the likelihood of choosing the 3 horses is 1/56 or .0179.

When using permutations and combinations in probability, you usually are attempting to determine the probability of some event against the probability of that event occurring by chance. Permutations and combinations give you all of the possible chance alternatives.

There are many cases of combinations and permutations that are of particular interest in statistics. One of them is the *binomial distribution,* which deals with combinations where there are only two possible outcomes: pass/fail, right/wrong, heads/tails, win/lose, etc. Many of the statistical problems dis-

• BOX 5.2

Chances of Winning the Perfecta by Chance in a Race with Eight Horses

WINNING TICKET NAMES THE FIRST, SECOND, AND THIRD PLACE HORSES

Formula: $$\text{prob}(w) = \frac{1}{_NP_M}$$ (5.3)

The probability of winning or prob(w) is 1 (for the one bet) divided by the total number of possible outcomes ($_NP_M$). The latter is obtained from the permutations formula.

STEPS

1. To find $_8P_3$, first set up the permutation formula.

$$_NP_M = \frac{N!}{(N - M)!}$$

$$_8P_3 = \frac{8!}{(8 - 3)!}$$

$$= \frac{8!}{5!}$$

2. Write out the factorials.

$$_8P_3 = \frac{8 \times 7 \times 6 \times 5 \times 4 \times 3 \times 2 \times 1}{5 \times 4 \times 3 \times 2 \times 1}$$

3. Cancel common elements.

$$_8P_3 = \frac{8 \times 7 \times 6 \times \cancel{5} \times \cancel{4} \times \cancel{3} \times \cancel{2} \times \cancel{1}}{\cancel{5} \times \cancel{4} \times \cancel{3} \times \cancel{2} \times \cancel{1}}$$

4. Multiply the remaining numbers.

$$_8P_3 = 8 \times 7 \times 6 = 336$$

5. Step 4 is the number of permutations. To find the probability of winning by chance, divide 1 by this number.

$$\frac{1}{336} = .0030$$

FOR COMBINATIONS: THE WINNING TICKET NAMES THE HORSES FINISHING IN THE TOP THREE

Formula:
$$\text{prob}(w) = \frac{1}{{}_NC_M} = \frac{1}{N!/M! \,(N-M)!} \qquad (5.4)$$

The probability of winning or prob(w) is 1 (for the one ticket) divided by the total number of possible outcomes (${}_NC_M$). The latter is obtained from the combinations formula.

STEPS

1. To find ${}_8C_3$, set up the combinations formula.

$$
\begin{aligned}
{}_NC_M &= \frac{N!}{M!\,(N-M)!} \\
&= \frac{8!}{3!\,(8-3)!} \\
&= \frac{8!}{3!\,5!}
\end{aligned}
$$

2. Write out the factorials.

$$
{}_8C_3 = \frac{8 \times 7 \times 6 \times 5 \times 4 \times 3 \times 2 \times 1}{(3 \times 2 \times 1)\,(5 \times 4 \times 3 \times 2 \times 1)}
$$

3. Cancel common elements.

$$
{}_8C_3 = \frac{8 \times 7 \times 6 \times \cancel{5} \times \cancel{4} \times \cancel{3} \times \cancel{2} \times \cancel{1}}{(3 \times 2 \times 1)(\cancel{5} \times \cancel{4} \times \cancel{3} \times \cancel{2} \times \cancel{1})}
$$

4. Multiply the remaining values.

$$
\begin{aligned}
{}_8C_3 &= \frac{8 \times 7 \times 6}{3 \times 2 \times 1} \\
&= \frac{336}{6} \\
&= 56
\end{aligned}
$$

5. Step 4 is the number of combinations. To find the probability of winning by chance, divide 1 by this number.

$$
\frac{1}{56} = .0179
$$

cussed later in the book are examples of binomial distribution. For the time being, let's consider some simple examples.

Consider the probability of a newborn child being either male or female. Suppose that you are at the delivery room of a hospital on New Year's Day. There are 6 deliveries that day. The nurse thinks that it is very unusual that of the 6 cases, there were 5 girls and 1 boy. What is the probability of this happening? To determine this we first consider that the probability of having a boy on any occasion is .5. The probability of having 2 boys born in succession is

$$.5 \times .5 = .25.$$

The probability of having three boys in succession is

$$.5 \times .5 \times .5 = .125.$$

In fact, for each consecutive birth, the probability of any specific order in 6 births is

$$.5 \times .5 \times .5 \times .5 \times .5 \times .5 = .0156.$$

This is a problem in combinations. Before giving the formula for calculating the combination, let's look at all of the ways we could get 5 girls in 6 consecutive births. In the following, B = boy, and G = girl.

```
B  G  G  G  G  G
G  G  B  G  G  G
G  G  G  B  G  G
G  G  G  G  B  G
G  G  G  G  G  B
```

As you can see, there are 6 different ways to have 5 girls born in 6 consecutive births.

Now let's use the combinations formula to analyze the same problem. Since we are looking for 1 boy in 6 births, we use the notation and formula shown in Box 5.3. To determine the probability of this event, we must divide the 6 combinations by all of the possible combinations of boys and girls in 6 births. As we noted above, the probability of a boy in any birth is .5. For this example we will express the .5 as 1/2. So, in 6 consecutive births, the probabilities of having any specific combination (that is, 5 girls, 1 boy) are expressed as $1/2 \times 1/2 \times 1/2 \times 1/2 \times 1/2 \times 1/2 = 1/64$. Now we can examine the chances of 5 female births in 6 consecutive births in the hospital. There are 6

● BOX 5.3

Number of Combinations of Having Only One Boy in Six Births

Formula: $${}_M C_N = \frac{N!}{M! \, (N - M)!}$$

$$
\begin{aligned}
{}_6 C_1 &= \frac{6!}{1! \, (6 - 1)!} \\[2mm]
&= \frac{6!}{1! \, (5)!} \\[2mm]
&= \frac{6 \times 5 \times 4 \times 3 \times 2 \times 1}{(1)5 \times 4 \times 3 \times 2 \times 1} \\[2mm]
&= \frac{6 \times \cancel{5} \times \cancel{4} \times \cancel{3} \times \cancel{2} \times \cancel{1}}{(1)(\cancel{5} \times \cancel{4} \times \cancel{3} \times \cancel{2} \times \cancel{1})} \\[2mm]
&= \frac{6}{1} \\[2mm]
&= 6.
\end{aligned}
$$

● BOX 5.4

What are the Chances of Having Three Boys Born in Eight Consecutive Births?

STEPS

1. Determine the number of possible unique sequences in 8 consecutive births.

Birth 1		Birth 2		Birth 3		Birth 4		Birth 5		Birth 6		Birth 7		Birth 8
$\frac{1}{2}$	$\times$	$\frac{1}{2}$	$\times$	$\frac{1}{2}$	$\times$	$\frac{1}{2}$	$\times$	$\frac{1}{2}$	$\times$	$\frac{1}{2}$	$\times$	$\frac{1}{2}$	$\times$	$\frac{1}{2}$

Since the probability of a boy is 1/2 for each birth, the number of possible outcomes is

$$\left(\frac{1}{2}\right)^8 = \frac{1}{256},$$

or there are $2^8 = 256$ possible unique sequences.

2. Define the combinations formula for the number of ways that 3 boys could be born in 8 births.

$$_8C_3 = \frac{8!}{3!\,(8-3)!}$$

3. Reduce the formula to

$$_8C_3 = \frac{8!}{3!\,(5)!}.$$

different ways of this happening and 64 potential outcomes of any type from 6 consecutive births. Therefore, the probability of 5 consecutive females in 6 births is

$6/64 = .0937$.

Box 5.4 shows the steps for calculating the probability of having 3 boys born in 8 consecutive births.

Summary

In this chapter, the concepts of permutations, combinations, and probability have been introduced. This has been a basic chapter to give you some under-

4. Expand and cancel.

$$_8C_3 = \frac{8 \times 7 \times 6 \times \cancel{5} \times \cancel{4} \times \cancel{3} \times \cancel{2} \times \cancel{1}}{(3 \times 2 \times 1)(\cancel{5} \times \cancel{4} \times \cancel{3} \times \cancel{2} \times \cancel{1})}$$

$$= \frac{8 \times 7 \times 6}{3 \times 2 \times 1} = \frac{336}{6}$$

$$= 56$$

As you come to understand the method, you should be able to cancel terms without physically writing them out. So you might move directly to the step

$$\frac{8!}{3! \, (5)!} = \frac{8 \times 7 \times 6}{3 \times 2 \times 1} = \frac{8 \times 7 \times \cancel{6}}{\cancel{6}}$$

$$= 56.$$

5. Step 4 shows that there are 56 different ways for 3 boys to be born in 8 births. Step 1 shows that there are 256 unique outcomes in 8 consecutive births. So the probability of having 3 boys born in 8 consecutive births is

$$\frac{56}{256} = .2188.$$

Stated another way, this event would be expected about 22 times in each sequence of 100 births.

standing of how to determine probabilities. The remainder of the book is devoted to the use of statistical tests. These tests are designed to relate observations to probability distributions. Statistical inference is designed to determine whether events we observe are attributable to things we understand or if they should be attributed to chance.

We live in a probabilistic world. To understand the complexities of the world we need to consider the chances of various events. The study of probability requires an understanding of random experiments in which the results are determined completely by chance. In the study of probability we attempt to determine how likely specific events are by chance alone.

The calculation of probability for independent events is aided by some basic rules. If we calculate the probability of one event *or* the other we use an additive rule. According to the additive rule we add the two probabilities. Sometimes we want to find the joint probability of two events. If we want to know the chances of obtaining a specific outcome for event 1 *and* a specific outcome for event 2, we use the multiplicative rule. According to this rule independent probabilities of each event are multiplied to find a joint probability.

The calculation of probabilities also requires formal methods of counting known as permutations and combinations. Permutations require predicting the precise order of an outcome; combinations do not consider order. Inferential statistics are used to determine the probability that certain observations could have occurred by chance alone. Throughout the rest of this book, we use basic concepts from probability theory repeatedly.

Key Terms

1. **Random Experiment**   An experiment in which the results are determined completely by chance.
2. **Set**   A collection of things or objects that are clearly defined by some rule.
3. **Element**   Any member within a set.
4. **Empty Set**   A set with no elements.
5. **Union**   For two sets *A* and *B*, the union is all elements that are in *A*, in *B*, or in both *A* and *B*.
6. **Intersection**   For sets *A* and *B*, the intersection is the subset of all elements that are in both *A* and *B*, but not in *A* alone or in *B* alone.
7. **Mutually Exclusive Events**   Two events are mutually exclusive if they share no common elements.
8. **Complement**   If there is a subsct from some defined set, its complement is made up of all othcr clcments in the set outside of the subset.
9. **Permutation**   A method of counting that requires the specification of the precise order in combinations of events.
10. **Combination**   A method of counting that does not require the specification of order for combinations of events.

Exercises

1. Bill Parsons is a junior at Ohio State University. In set theory, Bill would be referred to as
    a. a set.
    b. an element.
    c. a union.
    d. an intersection.

2. In the exercise above, students at Ohio State University might be considered

   a. a set.
   b. an element.
   c. a union.
   d. an intersection.

3. Draw a Venn diagram in which the intersection of two events P and Q is relatively large. Now draw another one in which the intersection is relatively small.

4. How many unique sequences would there be from 7 independent coin tosses?

5. You are conducting an experiment for a drug company in which rats will be given 3 different drugs. The company is evaluating 5 new drugs, but it would be too stressful to give animals all 5 drugs. Therefore, each animal will receive 3. However, the order in which the drugs are presented is very important. How many possible treatment combinations must you consider? Suppose you need 5 animals for each of these. How many animals will you need to perform the experiment?

6. At the beginning of the NFL football season, you would like to predict which teams will be in the playoffs. There are 28 teams in the league and 10 of them gain a playoff berth. What is the probability of selecting the playoff teams by chance? What is the probability of selecting 7 of the 10 playoff teams by chance?

6

Introduction to Inferential Statistics

Believe it or not. That was the challenge put to us in Mr. Ripley's famous series of books. Yet the challenge "believe it or not" extends far beyond the series of odd events described by Mr. Ripley. Throughout life we are confronted with thousands of facts others would like us to believe. Information is often confusing and contradictory, and assessing the validity of the information is very difficult. The remainder of this book is devoted to methods statisticians and scientists have developed to help them make consistent decisions about the meaning of evidence obtained in research studies.

The media often present results of scientific studies as very clear cut. The treatment is described as one that "works." We almost assume that it works in all cases and that its effects are unquestionably distinguishable from the effects of other treatments or of no treatment at all. Unfortunately, this is rarely the case in the behavioral and biological sciences.

Consider an experiment on how a new arthritis drug affects symptoms and activities of daily living. One group of arthritis patients gets the new drug while the others get a placebo, an inert substance that is known not to have pharmacological effects. The two groups are the same in daily function in the beginning of the study, but after two months the mean daily functioning score for the drug group is slightly higher than the mean for the placebo group. Yet upon inspection of the data, we find that some of the people in the placebo group improved more than some of the people in the drug group (see Table 6.1). In other words, on the average the people in the drug group improved

Table 6.1						
A hypothetical study of the effects of a new arthritis drug upon daily functioning: Does the drug work?						
DRUG GROUP				PLACEBO GROUP		
PATIENT	FUNCTION			PATIENT	FUNCTION	
	Before Treat-ment	After Treat-ment-	Change		Before Treat-ment	After Treat-ment	Change
John Jones	5	6	+1	Herman House	3	3	0
Sally Smith	4	6	+2	Elsie Evans	4	4	0
Eli Ellis	3	4	+1	Hank Hart	5	8	+3
Bill Brown	5	5	0	Tess Tomas	5	6	+1
Martha Miller	5	4	−1	Karen Karr	6	4	−2
Betty Billings	6	8	+2	Sharon Stevens	5	5	0
			$\bar{X} = .83$ $S = 1.17$				$\bar{X} = .33$ $S = 1.63$

more in function, but the data do not reveal perfect correspondence between taking the drug and getting better. Shall we believe that the drug works?

The question of whether the drug works is very complicated, as you can see from Table 6.1. Four of the six people taking the drug improved while only two of the six taking the placebo improved. On the other hand, one of the people taking the drug did not improve and one got worse. Furthermore, the individual who improved the most was Hank Hart in the placebo group.

Data such as those in the table are not unusual. In fact most behavioral and biological experiments produce data that leave us with some uncertainty about the outcome. A variety of methods in inferential statistics allows us to put results of experiments into context by relating the outcomes to well-understood laws of probability. In the case of the hypothetical drug experiment, differences between the drug and the placebo are not statistically significant. In other words, the differences are no greater than we would expect by chance. In the next few chapters you will learn some of the methods for interpreting experimental results.

Decision Rules

A decision is a choice from among several alternatives. You make many decisions everyday: what shirt to wear, where to eat lunch, etc. As a consumer of information you make decisions about whether you believe things you read or hear. While driving to school you may hear a radio commercial that says, "Ernie's Burger Barn has the best hamburgers you've ever tasted." Having eaten at Ernie's you may reject the proposition. In other words, propositions stated as true by the ad (Ernie's burgers are better than any others) may be viewed as untrue given your previous experiences at Ernie's, and you decide not to believe the information.

Scientists must make decisions about the meaning of scientific observations. To enhance communications, scientists have devised rules to help them make these decisions in systematic ways. In other words, we hope to agree on the rules of the game so that different scientists confronted with the same information will reach the same conclusions. These rules are fairly consistent across many different social, behavioral, and biomedical sciences. Virtually all academic disciplines teach students to analyze information critically before accepting it as true. Our standards for accepting information should be very high, and we should not endorse statements as true unless we have substantial evidence. The actual methods for obtaining the evidence differ from discipline to discipline. The social, behavioral, and biomedical sciences are similar in that they all use statistical probability distributions to help put observations in a context. It is common in each of these sciences to draw inferences about populations on the basis of samples.

Samples and Populations

In Chapter 3, we introduced the concepts of populations and sampling. These are very important concepts that will be worthwhile to review again. A *population* is a set of all possible observations of a specific type. This set includes all possible objects, people, or events with the same definition. For example, the population of Canadian women includes every female holding Canadian citizenship. The population of full-time students at Ohio State University includes all students currently enrolled at Ohio State who meet the university's definition of a full-time student.

Populations can be large or small. For example, a population of dogs may be extremely large and may include all dogs in the world. Within that population, we may define another population of specific interest. This might be the population of Irish setters within the small town of Del Mar, California. For many problems in social, behavioral, and biomedical sciences, we want to make statements that apply to relatively large populations. For example, we like to say that a certain form of psychotherapy is effective, a particular drug works, or that social support has a specific influence. In other words, we wish to make inferences to very general populations. Therefore we tend to think of populations as large.

Some important characteristics of a population can be described by its mean and its standard deviation. Yet if the populations are large, we can't always make all the observations necessary to calculate these statistics. Calculating the mean height of American men, for example, requires that we measure each man and take the average. Fortunately there are methods for estimating the mean of a population without observing every single element. This is accomplished by taking from the population *samples* that are representative and unbiased.

A sample is a subset of observations drawn from the population. Ideally the sample should be drawn in a representative and unbiased manner. Samples help us estimate characteristics of the population. Generally speaking, the larger the sample, the more assured we are that the sample is representative of the population. For instance, if you wanted to estimate the height of American men, and you drew a sample of two men randomly from the population and measured their heights, how confident would you be that you had accurately captured the height of the average American man? Probably not very confident. Now suppose that you increase the number of observations to one hundred. How confident would you feel about your estimate in this case? Now consider a sample size of 4 million. Do you think that your confidence would increase?

Statisticians have worked out how large samples need to be to accurately represent the population means. As we show later, the sufficient sample size depends on characteristics of the variable under study, such as its variability. This is illustrated in Fig. 6.1. The figure shows a variety of samples drawn from the same population. The numbers show scores on a hypothetical variable. This population has 20 elements. One component of Fig. 6.1 shows a small sample of 2 observations. You can see that the mean of this sample is slightly different from the population mean. The second example shows 10 observations, and the mean is somewhat closer to the population mean. Finally the third sample has 18 elements. Here the sample is almost the same as the population. As a result, the discrepancy between the sample mean and the population mean is quite small.

Despite the advantages of having a large sample, largeness alone does not ensure accurate correspondence between a sample mean and a population mean. Suppose we are attempting to estimate the height of men. The population for this example would be all men in the world. To investigate this we draw a sample of 65 million American men. Would this very large sample of 65 million give us an accurate reading of the height of men? Probably not. American men may not be representative of all men in the world. For example, the average American man is taller than the average Japanese man. By taking a much smaller sample that is representative of the world population of men, the estimate of the population mean would most likely be more accurate. The classic example of nonrepresentative or biased sampling with a large sample size is illustrated in the case of the 1936 Literary Digest Poll in Box 6.1.

To ensure that the selection of a sample is unbiased, we must sample in a random fashion. A **random sample** is a sample in which each member of the

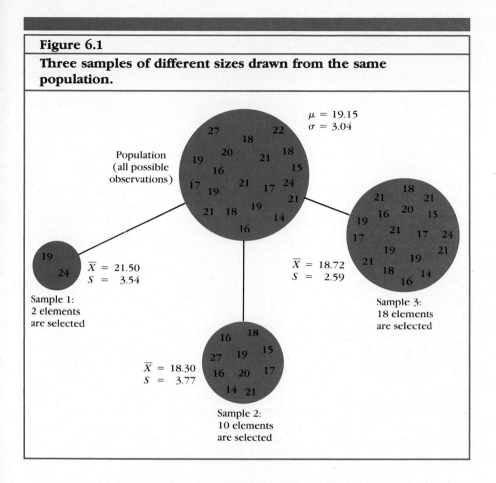

Figure 6.1

Three samples of different sizes drawn from the same population.

$\mu = 19.15$
$\sigma = 3.04$

Population (all possible observations)

27 18 22
20 18
19 21
16 21 15
17 21 17 24
19 21
21 18 19 14
16

$\overline{X} = 21.50$
$S = 3.54$

19
24

Sample 1:
2 elements are selected

$\overline{X} = 18.72$
$S = 2.59$

18
21 21
16 20 15
19 21 17 24
17 19 19 21
21 18 14
16

Sample 3:
18 elements are selected

$\overline{X} = 18.30$
$S = 3.77$

16 18
27 19 15
16 20 17
14 21

Sample 2:
10 elements are selected

population has an equal chance of being selected. (Methods of random sampling will be discussed in more detail later in this chapter.) In the example of a sample of heights for the world's population of men, each man in the world would have an equal probability of becoming a member of the sample. In a random sample of any size, each element is an unbiased sample from the population. The size of the sample can be as small as 1 and as large as the population minus 1. The unbiasedness is assured by random selection—giving every element in the set an equal chance of being drawn.

Although a very large sample may seem representative, people often think that it ensures that summary statistics such as the mean and standard deviation for the sample will approximate the corresponding values for the population. This is not necessarily so. In fact, statisticians have worked out the sample sizes required to estimate corresponding values for populations with known characteristics. For many situations it is possible to get away with a relatively small sample. The key is that the sample is drawn in a random, unbiased fashion.

● BOX 6.1

The Case of the Literary Digest Poll

A LARGE SAMPLE DOES NOT ALWAYS ENSURE ACCURACY

In a famous case, a magazine called the *Literary Digest* attempted to forecast the outcome of the 1936 presidential election between Roosevelt (the Democrat) and Landon (the Republican). The

Oh! Oh!

magazine drew its sample from its readers, from automobile registrations, and from telephone directories. In 1936, all of these sources overly represented the wealthy; most of them were Republicans. The poll showed that Landon would win by a landslide. The results of the election were, however, just the opposite. Roosevelt won by one of the greatest margins in American history. Thus, survey results are of little value if the sample is not random. Election day polls using as few as 2000 respondents to represent all of the voters in the United States have repeatedly been shown to be very accurate when the small samples are drawn randomly.

In 1936 Franklin D. Roosevelt won the presidential election by a landslide. A poll by the Literary Digest *with a very large sample size predicted that Roosevelt's Republican opponent Alfred Landon would win the election by a landslide. However, the sample for the study was not drawn in a random fashion, and Republicans had a much higher probability of being included in the sample.*

By Morris From George Matthew Adams Service

One example of a small sample estimating characteristics of a large population is the method used for television ratings. You have probably never been contacted by television pollsters, yet they estimate your television viewing habits every week. The Nielson Company, a commercial organization, places electronic devices on the television sets of a limited number of American families. The devices record which stations are watched at all hours of the day and night. On the basis of a very small sample of households (about 1200 nationwide), Nielson can make a fairly accurate estimate of the number of persons watching particular programs. In fact, other studies with larger samples have tended to support the validity of the Nielson ratings. Further, large businesses such as advertising agencies apparently believe the Nielson ratings. Sizable amounts of money for advertising fees are based on these ratings.

In summary, in drawing a sample we should make sure that it is random and representative. This assures that all people have an equal probability of being included. When the samples are not randomly selected, the results may be inaccurate.

Rules and Decisions

Science has many methods for seeking the truth. There are many forms of inquiry, and they each have different rules. Yet many different systems of human inquiry have well-structured and well-defined rules of evidence.

Consider methods for inquiry in criminal law. Two principles dominate in criminal proceedings:

1. The accused is innocent until proven guilty.
2. A conviction must be based on evidence that is beyond a reasonable doubt.

These principles, based on old English law, have stood the test of time. They have been an enduring part of English law and something that most students learn about in high school civics courses.

Scientific evidence is based on similar principles. Recall the experiment described in Table 6.1. A sample of arthritis patients was randomly divided into two groups and assigned to take either a drug or a placebo. If both groups had been given equivalent treatment we would assume that they are two samples from the same population. Each should therefore be the same. However, if the drug had some effect, those receiving the treatment would be expected to differ from those given the placebo. In other words, the groups would differ more than would be expected by chance.

In scientific inference we assume that groups do not differ until there is substantial evidence that they are not the same. The assumption that the groups do not differ is called the *null hypothesis*. This is a formal statement that there is no difference, no relationship, no correlation, etc., among groups. If we think back to the analogy of legal reasoning, the null hypothesis is similar to the notion of innocent until proven guilty. We assume that a drug is no more effective than a placebo until we prove with evidence that it makes a difference.

But under what circumstances is the evidence strong enough for us to reject the null hypothesis? Again there is a similarity to the legal system. In criminal law we say that we assume the accused is innocent until we have evidence beyond a reasonable doubt. Similarly in statistical hypothesis testing we assume that the null hypothesis is correct until we have evidence beyond a reasonable doubt that it is incorrect. Under these circumstances, we reject the null hypothesis in favor of an *alternative hypothesis* that states that the observed differences are large enough that it is unlikely they occurred by chance. In the next section we develop in more detail the concepts of null hypothesis, alternative hypothesis, and the formal test between them.

■ Sampling from the Same Population

When teaching introductory statistics, I often write 200 numbers on small pieces of paper and place them in a large shoe box. The 200 numbers represent a normal bell-shaped distribution. Then I have students draw random samples of 20 from the shoe box, and we calculate the mean and standard deviation for each sample. After these statistics have been computed, the sample numbers are returned to the box. Since the samples are random and unbiased, the mean of each sample is expected to be an unbiased estimate of the population mean. Should the sample means be the same as one another? (Another version of this experiment is described in Box 6.4.)

If you perform this experiment, and I encourage you to do so, you will probably find that the means of each sample of 20 are not the same. The reason for this is that each sample mean is an *unbiased estimate* of the population mean. Yet these sample means can and do differ from one another. So each time this experiment is performed, the sample means differ from one another. In statistical analysis the task is often to estimate whether observed differences between sample means can be attributed to sampling error or to some systematic source of variation.

If samples are drawn from the same population, we expect that sample means will not all be the same. Statisticians have worked out the probability that differences between observed means could be obtained by chance alone. These methods are central to inferential statistics.

Parameters and Statistics

Before proceeding, it may be worthwhile to consider some important differences between the symbols and terminologies used to refer to samples and populations. This distinction was mentioned briefly in Chapter 3, but it is important to review it in more detail here.

■ Populations

As noted above, populations include every member of a defined class. The mean and the standard deviation of a population are referred to as *parameters*. Typically, we signify these parameters with Greek letters. The Greek letter mu μ is used for mean, and Greek letter sigma σ is used for the standard deviation.

Table 6.2		
Terms and symbols used to describe populations and samples		
DEFINITION	POPULATION	SAMPLE
	All Elements with Same Definition	A Subset of the Population Usually Drawn to Represent it in an Unbiased Fashion
Descriptive characteristics	Parameters	Statistics
Symbols used to describe	Greek	Roman
Symbol for mean	μ	$\bar{X}$
Symbol for standard deviation	σ	S

■ Samples

A subset of the population is a *sample*. Means, standard deviations, and other values that describe characteristics of the sample are known as *statistics*. In describing statistics, we use Roman letters. Mean is usually denoted by $\bar{X}$, and the standard deviation by S. Table 6.2 summarizes some of the terminology for populations and samples.

Sampling Distributions

In many applications of statistics, we are really interested in population parameters. However, in the real world, finding population parameters is usually beyond our means. For instance, the United States government does perform a census every 10 years. In the census, information is obtained about every single individual in the country. Yet the census costs millions of dollars. Much of the census information could be estimated with considerably smaller samples. For instance, nationwide public opinion polls accurately forecast the outcome of elections with as few as 2,000 respondents. In many cases the additional expense required to find population parameters is not justifiable, since we can obtain similar information by drawing unbiased samples.

Most times, then, the task is to estimate population parameters using statistics. Theoretically, our sights are set on estimating a population parameter such as the mean. To estimate the population mean, we draw a series of samples, calculate the sample means, and use the average of the sample means as the estimated population mean.

Methods for obtaining sample means are important. There are two ways to draw a random sample: with and without replacement. When we draw a sample without replacement, we do not put it back in the pool once it has been drawn. For example, suppose you had a hat with 100 numbers in it. You draw 3 numbers from the hat: 7, 11, and 8. After the numbers are drawn, they remain out of the hat. Now suppose that you perform the same drawing, but each time a number is drawn and recorded, it is placed back in the hat so that it has the possibility of being drawn again in future trials. This is called sampling with replacement. For most problems in inferential statistics, we assume that sampling is done *with replacement*.

If we were to draw sample means from a population of, say, 200 observations *with replacement*, we could have an infinite number of sample means. Since replacement puts numbers back into the hat after they have been drawn, and numbers are drawn one at a time, we never run out of numbers in the hat.

Remember that each time we draw a sample from the hat (or population) and calculate the sample mean, we assume that it is an unbiased estimate of the population mean. As we continue to draw sample means, we can make a frequency distribution of sample means. This distribution is known as a **sampling distribution.** A sampling distribution is defined as the theoretical probability distribution of values that could be obtained from some statistic in random samples where a particular sample size is taken from a population (Yaremko et al., 1982).

If you create a sampling distribution, you will be able to calculate its mean as well as its standard deviation. The mean of the sampling distribution is an unbiased estimate of the mean of the population. However, as we look at the sampling distribution (Box 6.2) we see that not all sample means are at the mean of the sampling distribution. Some are above the mean and some are below it. Yet given that there is a large enough number of sample means, they will tend to be normally distributed around the mean of the sampling distribution. The difference between any sample mean and the mean of the sampling distribution is known as **sampling error.** Sampling error is random and we would expect that the magnitude of sampling error above the mean is equal to the magnitude of sampling error below the mean. In other words, sampling errors should balance out around the mean in a normal distribution. Box 6.2 illustrates this concept.

As a rule, the variability of sample means around the mean of the sampling distribution decreases as the sample size increases. Consider the estimate of the mean of a population of 150 when each sample mean is based on 1 observation, 10 observations, 50 observations, and 100 observations. As the number of observations in the sample increases, the variability of the sample mean around the population mean decreases. Large samples include more individual estimates of the population mean and indeed include more of the values in the population. This is illustrated in Box 6.3.

In practice we rarely take repeated random samples from the same population. This would simply be too much effort. Instead we usually take just one sample, and we want to know how well this single sample represents the population parameters. Thus, most of our methods are designed to help us determine how well we are doing with a single sample.

As was shown in Box 6.3, we are more likely to accurately estimate the population mean if we take larger samples. In fact, the difference between the sample mean and the population mean decreases as the number of observations used to calculate the sample mean increases. Although each observation is an estimate of the population mean, averaging larger numbers of these esti-

● BOX 6.2

Distribution of Sample Means Around a Population Mean

Suppose that we want to estimate the average birth weight of babies born in a large university hospital. Records of births over the course of 3 years are obtained and random samples of 20 births are drawn. Each time a sample is drawn, a mean is calculated. This process is repeated 400 times. Then a frequency distribution of the sample means is created. (See Fig. 6.2.)

The mean of the sampling distribution of sample means is 3300 gm. This is an unbiased estimate of the population mean.

Now consider the mean of a particular sample of 20 births that was 3150 gm. The difference between this sample mean and the population mean is shown in Fig. 6.3.

The difference between the population mean and the sample mean is the sampling error, which in this case is −150 gm. According to sampling theory, for each sample with a negative sampling error, there is another sample with an equally large positive sampling error. Thus, over an infinite number of samples, the sampling error cancels out.

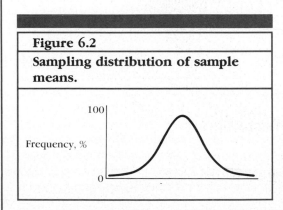

Figure 6.2

Sampling distribution of sample means.

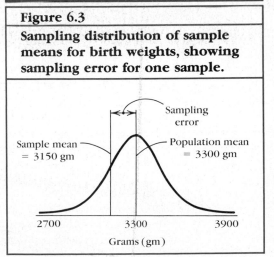

Figure 6.3

Sampling distribution of sample means for birth weights, showing sampling error for one sample.

● BOX 6.3

Sample Size and Sampling Error

As the sample size increases, the sampling error decreases. Sampling error is estimated from the variability of sample means around a population mean.

Consider sampling from a population of 1,000, using samples of 900, 500, 100, and 1. Let's represent the population as a circle. Now we can represent the different sample sizes as portions of the population. (See Fig. 6.4.) As you can see, the larger the sample, the more of the population is represented.

For large, unbiased samples, the sample means tend to be close to the population mean and may not vary much from one another.

As sample size decreases, there is greater variability around the population mean. This is illustrated in the Fig. 6.5.

The figure shows relatively less variability around the population mean for the large sample. As sample size decreases, there is greater variability in the distribution of sample means around the population mean.

Figure 6.4

Relationship of samples of various sizes to population of 1000.

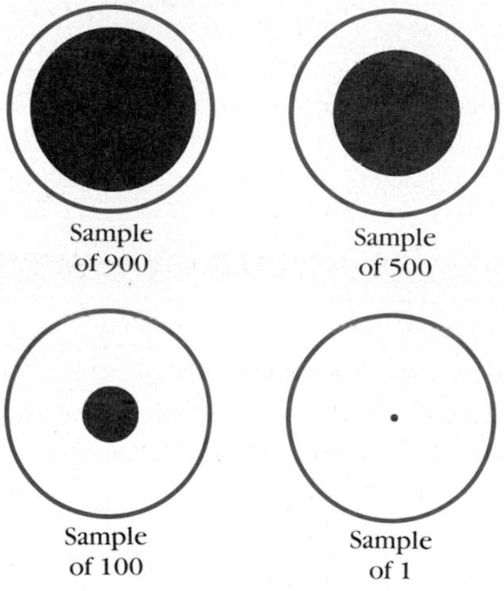

Sample of 900

Sample of 500

Sample of 100

Sample of 1

Figure 6.5

Relationship between sample size and variability about population mean for sampling distributions of sample means.

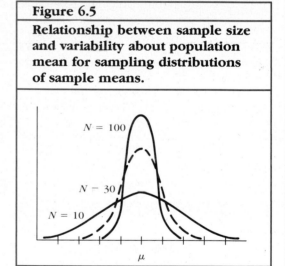

$N = 100$

$N - 30$

$N = 10$

μ

mates provides greater accuracy. Although repeated random samples from the same population do not give the same mean, the chances of closely approximating the population mean are better with a large rather than a small random sample. An exercise on drawing sample means from a population is given in Box 6.4.

● BOX 6.4

Taking Random Samples from a Population

This exercise demonstrates that repeated samples from the same population do not have exactly the same mean and standard deviation.

STEPS

1. Write numbers on 200 small pieces of paper and drop them into a hat. Try to choose numbers that might represent a normal distribution. For example, you might use the distribution of heights of men (in inches) and use examples of people at your university.
2. Shake the numbers inside the box or hat very well. Then randomly draw 20 numbers. Each time a number is drawn, write it down and place it back into the hat. This is sampling with replacement (see the text).
3. Calculate the mean and standard deviation for the sample.
4. Now mix the numbers again and draw a second sample.
5. Calculate the mean and standard deviation for the second sample.
6. Repeat the same procedure for a third sample.

Did all three samples have the same mean? If you were to draw a large number of samples from this population and calculate the mean for each, their distribution would be the sampling distribution of sample means.

Standard Error of the Mean

Now here is our dilemma: we need to know the population mean, but finding it would be very expensive because we have to measure every member of the population. Instead, we use a sample mean to estimate the population mean, although we know that the sample mean may not be exactly the same as the population mean. We could take repeated random samples of a particular size and use them to create a sampling distribution. The mean of this sampling distribution would be an unbiased estimate of the population mean. Yet, taking repeated random samples may be as difficult and as expensive as measuring every member of the population. Thus, in the real world we most often only have one sample mean.

The first step in estimating how well a sample mean (which is a statistic) represents a population mean (which is a parameter) is to find the **standard error of the mean.** This is an estimate of the standard deviation of the sampling distribution. (Recall that the sampling distribution is a frequency distribution of an infinite number of sample means. The mean of the sampling distribution estimates the mean of the population.) The standard deviation of this distribution is the standard error of the mean.

Using each value in the sample as an unbiased estimate of the population mean we can calculate a standard deviation for the sample; this serves as an estimate of the population standard deviation. The standard error of the mean is equal to the population standard deviation divided by the square root of the sample size. The formula for the standard error of the mean is

$$\sigma_{\bar{x}} = \frac{\sigma}{\sqrt{N}}. \tag{6.1}$$

(See Equation 3.5 for the definition of σ.)

Notice that the formula uses the Greek sigma for the standard deviation. Theoretically, the population standard deviation should be used in the calculation of the standard error of the mean; however, in practice we rarely know the population standard deviation, so instead we use the sample standard deviation. The formula becomes

$$S_{\bar{x}} = \frac{S}{\sqrt{N}}. \tag{6.1a}$$

(See Equation 3.7 for the definition of S.)

Take the example of the standard deviation of homicide rates in 10 states shown in Box 3.8. That example assumed that the 10 states were a population, but it is more appropriate to think of those 10 states as a sample from the population of 50 states. The standard deviation of the homicide rates was 4.2. The sample size N was 10. Placing these numbers in the formula, we find that the standard error of the mean is

$$S_{\bar{x}} = \frac{4.2}{\sqrt{10}} = 1.33.$$

Box 6.5 is an example for calculating the standard error of the mean.

Let's take two more examples using different standard deviations and sample sizes. First consider a sample of 27 with a standard deviation of 6.1. Plugging these values into the formula for the standard error of the mean gives

$$S_{\bar{x}} = \frac{6.1}{\sqrt{27}} = 1.17.$$

Now consider a sample with a sample size of 60 and a *variance* of 107.4. How would you place this into the formula? Remember that the variance is the standard deviation squared, so we must first change the variance of 107.4 to a standard deviation by taking its square root.

$$\sqrt{S^2} = S$$
$$\sqrt{107.4} = 10.36$$

Now we are ready to put the values into the formula. The standard error of the mean is

$$S_{\bar{X}} = \frac{10.36}{\sqrt{60}} = 1.34.$$

Actually, we could have divided 107.4 by 60 and taken the square root of that ratio to obtain the same value. This tells us that an alternative expression

● BOX 6.5

Calculation of the Standard Error of the Mean

GAMES WON BY 10 PROFESSIONAL TEAMS DURING REGULAR PLAY IN THE 1982 SEASON (THE SEASON SHORTENED BY A PLAYERS' STRIKE).

Redskins	8
Raiders	8
Dolphins	7
Chargers	6
Packers	5
Browns	4
Lions	4
Seahawks	4
Rams	0
Colts	0

STEPS

1. Calculate the standard deviation using the formula

$$S = \sqrt{\frac{\Sigma X^2 - \frac{(\Sigma X)^2}{N}}{N - 1}}. \qquad (3.8)$$

$$S = 2.888$$

(If you have difficulty using this formula, review Box 3.8 in Chapter 3.)

2. Find $\sqrt{N} = \sqrt{10} = 3.16$.
3. Put the results of Steps 1 and 2 into the formula for $S_{\bar{X}}$.

$$S_{\bar{X}} = \frac{S}{\sqrt{N}}$$

$$= \frac{2.888}{3.16}$$

$$= .91$$

of the standard error of the mean is the square root of the ratio of the variance over the sample size, or

$$S_{\bar{X}} = \sqrt{\frac{S^2}{N}}. \tag{6.2}$$

The standard error of the mean helps us understand the probability that a sample mean differs from the mean of a sampling distribution. It is worth reconsidering these concepts. Remember that the major task in inferential statistics is to estimate the population mean on the basis of sample means. If infinite repeated random samples are taken from the population, the mean of all such samples is equal to the population mean. However, it is not practical to draw an infinite number of samples to estimate the population mean. Therefore, a smaller number of samples is used, but the mean of these sample means may not be exactly the same as the population mean.

The difference between the sample mean and the population mean is called sampling error. The standard error of the mean is an estimate of the variability of sample means around the population mean. It is found by dividing the standard deviation of sample means by the square root of the number of samples.

Often only a single sample is drawn. In this case, each member of the sample is considered an unbiased estimate of the population mean. The standard deviation of the sample is taken as an unbiased estimate of the population standard deviation. The sample standard deviation divided by the square root of the sample size is an estimate of the standard error of the mean. The standard error of the mean helps us determine how closely the sample mean represents the population mean.

Confidence Intervals and Confidence Limits

Finding the standard error of the mean is one step toward estimating the population mean on the basis of the sample statistic. Making this type of estimate is an everyday occurrence for statisticians and scientists. We really want to know the population mean, yet we only have one sample from which to estimate. How do we know that the sample mean is the same as the population mean? The answer is that we do not know. In fact, it is unlikely that the sample mean will ever be exactly the same as the population mean. How, then, can we possibly use one sample as an estimate of a population parameter? We can use statistical inference to estimate the probability that the population mean falls within a defined interval. Since sample means are distributed normally around the population mean, we know that the sample mean is most probably near the population value. However, it is possible that the sample mean is an overestimate or an underestimate of the population mean. Given what we know about the standard deviation of the sampling distribution (the standard error of the mean), we can place the observed differences between the sample and population means into context.

The ranges that are likely to capture the population mean are called **confidence intervals.** Confidence intervals are bounded by **confidence limits.** A confidence interval is defined as a range of values with a specified probability of including the population mean. A confidence interval is typically associated with a certain probability level. For example, the 95% confidence interval has a 95% chance of including the population mean. A 99% confidence interval is expected to capture the true mean in 99 of each 100 cases. *Confidence limits* are defined as the values or points that bound the confidence interval.

The task of defining this interval requires that we first obtain a sample mean and then define an interval around it that most probably captures the population mean. We use the following formula to calculate the confidence interval:

$$CI = \overline{X} \pm Z_\alpha S_{\overline{X}}. \tag{6.3}$$

Let's go through the different components of this formula. $\overline{X}$ is the sample mean. Plus or minus ($\pm$) means we add to get the upper boundary of the interval (or upper limit) and subtract to get the lower boundary (or lower limit). Z_α is the Z-score for a certain probability (see Equation 4.2), and $S_{\overline{X}}$ is the standard error of the mean.

Suppose that we draw a random sample of children participating in a special early childhood education program in British Columbia and measure their IQs. We find that the mean IQ for the children in the sample is 108.78, and the standard deviation is 10.31. We would like to know the population mean for all children in this special program. Of course we cannot know this without measuring each child in the population, but we can state with a known probability the likelihood that the population mean falls within a defined interval around 108.78.

Box 6.6 shows the calculation of a 95% confidence interval for this example. The first step is calculating the standard error of the mean. In this case, N is 100, so

$$S_{\overline{X}} = \frac{10.31}{\sqrt{100}} = 1.03.$$

The next step involves deciding Z_α. In this example, we want to create a 95% confidence interval. Figure 6.6 shows the interval that we wish to create around the sample mean. The 95% interval is the nonshaded area in the figure. So we are cutting off about 2.5 percentage points at the top of the distribution and 2.5 percentage points at the bottom of the distribution. To find Z_α, we need the Z-score for the 97.5 percentile and the 2.5 percentile.

● BOX 6.6

Creating a 95% Confidence Interval

THE IQs OF CHILDREN IN A SPECIAL EARLY EDUCATION PROJECT

Data:

$$\text{Sample mean } \overline{X} = 108.78$$
$$\text{Sample standard deviation } S = 10.31$$
$$N = 100$$

Formula: $CI = \overline{X} \pm Z_{\alpha}S_{\overline{X}}$

STEPS

1. Find the standard error of the mean.

$$S_{\overline{X}} = \frac{S}{\sqrt{N}}$$
$$= \frac{10.31}{\sqrt{100}}$$
$$= \frac{10.31}{10}$$
$$= 1.03$$

2. Set Z_{α}. For the 95% level, it is 1.96 (see Appendix 1).
3. Obtain the product $Z_{\alpha}S_{\overline{X}}$ by multiplying the results of Step 1 and Step 2.

$$Z_{\alpha}S_{\overline{X}} = 1.96 \times 1.03 = 2.02$$

4. Obtain the lower confidence limit by subtracting the results of Step 3 from the mean.

$$\text{Lower limit} = \overline{X} - Z_{\alpha}S_{\overline{X}} = 108.78 - 2.02$$
$$= 106.76$$

5. Obtain the upper confidence limit by adding the results of Step 3 to the mean.

$$\text{Upper Limit} = \overline{X} + Z_{\alpha}S_{\overline{X}} = 108.78 + 2.02$$
$$= 110.80$$

6. State the confidence interval using the confidence limits obtained in Steps 4 and 5.

Statement: The probability is 95 in 100 that the interval from 106.76 and 110.80 includes the population mean.

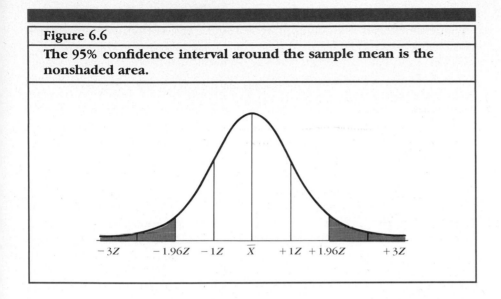

Figure 6.6

The 95% confidence interval around the sample mean is the nonshaded area.

To find the Z-values, turn to Appendix 1. The Z-score for the 97.5th percentile can be found by locating the column labeled "Area from −∞ to Z." The symbol ∞ means infinity. So −∞ suggests that the interval would include any score below Z, regardless of how low. Go down that column until you find .9750. This suggests that 97.5% of the cases fall below this point. Now move your finger left across the page until you are under the column labeled "Z." The Z-value for the row that contains .9750 is 1.96. The Z-value for the 2.5th percentile is a little trickier to find. The fourth column in Appendix 1 is labeled "Area from Z to +∞." Actually, you want to find the area from −∞ to Z. However, since the Z-distribution is symmetrical, the area between −∞ and Z is equal to the area between Z and +∞. Since the Z-score is below the mean, we multiply by −1. Using Appendix 1, we find that a Z of 1.96 is associated with an area between Z and ∞ of .0250. To get the Z-score for the 2.5th percentile, we multiply the Z-score by −1 to obtain −1.96. In this example Z_α is 1.96.

Remember from Chapter 4 that Z-scores are in standard deviation units. The standard deviation for a sampling distribution is the standard error of the mean. Thus, the probability level can be converted into standardized units by multiplying the Z_α-value by the standard error of the mean $Z_\alpha\sigma_{\bar{x}}$). Whenever we say Z_α, the α (alpha) refers to the probability level with which we are dealing. For example, if we had chosen a 99% confidence interval, we would have needed to choose a Z-value that leaves only 1/2 percent in each tail in the distribution. This Z-level is 2.58. The 95% and the 99% confidence intervals are the ones used most frequently in sample statistics.

As shown in Box 6.6, the next step involves subtracting the product of Z_α and $S_{\bar{x}}$ from the sample mean to find the lower limit of the confidence interval.

Then, we add this same value to the mean to obtain the upper limit of the confidence interval. The values representing the lower and the upper limits of the confidence interval are called the confidence limits.

Once the confidence interval has been calculated, we can make a statement about the population mean even though we have not observed it. We can state that the probability is 95 in 100 that the population mean falls between the confidence limits. For the example of IQ scores of children in the special program in British Columbia, we have drawn a sample of 100 children and tested them. Their mean IQ score was 108.78. The confidence level that we created has a lower limit of 106.76 and an upper limit of 110.80. So although we do not know the true population mean, we can say that the probability is 95 in 100 that the mean falls within the interval bounded by these two values.

Let's take one more example. Suppose that a new diet has come out and has been widely publicized across the country. We know that thousands of people are following the diet, and we would like to know how much weight they are losing. It is not feasible to weigh every single person who is using the diet, so we select a random sample of people across the country who claim to be using the diet, and we weigh them on several occasions. We find that over the course of one month, the mean weight loss on the diet is 5 pounds with a standard deviation of 6 pounds. There are 250 people in this sample. We would like to say that the population mean falls within a certain interval with 99% confidence.

The first step is to find the standard error of the mean.

$$S_{\bar{X}} = \frac{S}{\sqrt{N}} = \frac{6}{\sqrt{250}} = \frac{6}{15.81} = .38$$

The next step is to find Z_α. In this case, alpha is for the 99% confidence limit. The 99% confidence interval includes 49.5% of the area above the mean and 49.5% of the area below the mean. So you must go to Appendix 1 and find the value of the Z-score associated with the probability 49.5% above and below the mean. This value is 2.58 (and -2.58).

Next take the product of Z_α and $S_{\bar{X}}$. This is $2.58 \times .38 = .98$. Now we are ready to find the lower and upper confidence limits. The lower limit is obtained by subtracting .98 from the mean of 5 pounds.

Lower Limit $= 5 - .98 = 4.02$.

Next we find the upper limit by adding .98 to the mean.

Upper Limit $= 5 + .98 = 5.98$.

So the 99% confidence interval is 4.02 to 5.98. With this information, we can state that the probability is 99 in 100 that the weight loss in the population falls between 4.02 and 5.98 pounds.

It is always important when making these statements to consider whether the sample was really a random and representative sample from the population. If it was not, the statement would not be tenable.

■ Smaller Samples

Many years ago, mathematicians proved that the sampling distribution of sample means is a normal distribution. This is true whether or not the population from which the samples are drawn is normally distributed. Many of the important assumptions made in statistical analysis require that the sampling distribution be normal, or bell-shaped. Statisticians have proved that as the sample size increases, the more likely it is that the sampling distribution will be normal. In other words, as a greater number of sampling units are included, the greater is the likelihood that the distribution will approach normality. Further, the distribution of sample means will be normal even though the population distribution may not be normal. This is known as the *Central Limit Theorem*. Although the derivation of the Central Limit Theorem is beyond the scope of this book, its validity is well established. A brief illustration of the Central Limit Theorem is given in Box 6.7.

So far we have talked only about drawing relatively large samples from populations and estimating population parameters on the basis of these observations. Yet many times, we draw small samples. Under these circumstances, we may be less confident that the sampling distribution of sample means is normal. With smaller samples, it is sometimes appropriate to reference our observations to other sampling distributions that take into consideration the small sample size. One example is the *t*-distribution, which is covered in detail in Chapter 7.

Placing observations into a frame of reference known as a sampling distribution helps us understand their meaning. If someone told you they had observed a lowland gorilla that weighed 400 pounds you may not know whether the gorilla was heavy or light without knowing more about the weights of gorillas. If you later discovered that the mean weight for a lowland male gorilla is 380 pounds with a standard deviation of 20 pounds, you could say that this gorilla was heavier than average. In other words, knowing about the distribution of weights of lowland gorillas helps you interpret the meaning of this particular observation. The standard normal distribution is a special case in which many different scales of measurement (weight, height, IQ) are converted into standard deviation units. This transforms any set of observations into one having a mean of 0 and a standard deviation of 1. Using Appendix 1 we can find that a variable with a Z-score of 1.0 is in the 84th percentile.

Although the Z- or standard distribution is of great value, there are situations in which it may not be appropriate. Specifically, it is not used when the population standard deviation is not known, and we must estimate it based on information obtained in a sample. In other words, when we are using S instead of σ. The other occasion for not using the Z-distribution is when we are making inferences on the basis of a small sample. Some of the assumptions about the normality of the sampling distribution are not appropriate for samples less than 30.

● BOX 6.7

Brief Illustration of the Central Limit Theorem

According to the Central Limit Theorem, the sampling distribution of sample means will tend toward normal. This happens whether or not the populations of observations that are sampled is normally distributed. Consider sampling from population distributions of three different shapes (see Fig. 6.7).

Repeated random samples of size 10 are taken from each population and a sampling distribution is created. The distributions are illustrated in Fig. 6.7.

Finally random samples of size 100 are drawn, sample means calculated, and the sampling distributions created.

As the example demonstrates, the sampling distribution of sample means rapidly approaches normality independent of the distribution of population scores. For sample sizes of 10 or greater, the sampling distribution of sample means can be taken as normal.

Figure 6.7
Three population distributions.

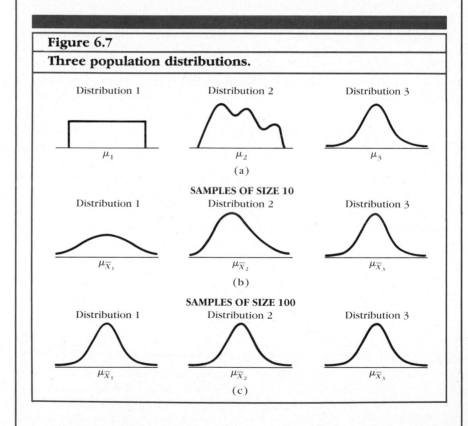

Without our old friend the *Z*-distribution, it is necessary to find other distributions that we can use to put our observations into context. In the next chapter we will focus on the *t*-distribution.

The *t*-distribution is not a single distribution but a family of distributions each with its own **degrees of freedom.** The degrees of freedom are defined as the sample size minus one, or $N - 1$. A technical discussion of the concept of degrees of freedom is given in Box 6.8. Each of the *t*-distributions takes into consideration the size of the sample. Figure 6.8 shows several *t*-distributions with different degrees of freedom. As you can see from the figure, the distribution is flatter with more area in the tails of the distribution when there are few degrees of freedom. With more degrees of freedom, the distribution becomes more peaked in the center with less area in the tails. A *t*-distribution with 30 degrees of freedom is already quite similar to the *Z*- or standard normal distribution. In Fig. 6.8 the *t*-distribution for infinite degrees of freedom is exactly the same as the *Z*-distribution.

Figure 6.8
t-distribution for various degrees of freedom.

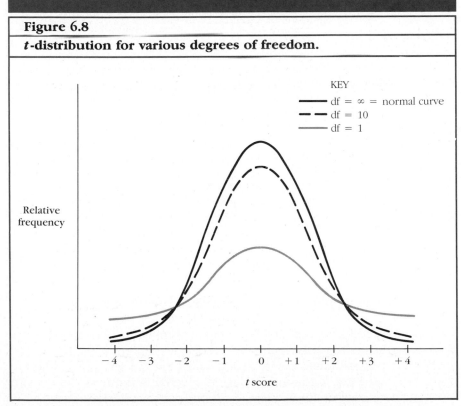

KEY
— df = ∞ = normal curve
– – df = 10
— df = 1

Source: Robert B. McCall, *Fundamental Statistics for Psychology, Second Edition.* Harcourt, Brace, Jovanovich, 1975. Reprinted by permission of the publisher.

● BOX 6.8

The Concept of Degrees of Freedom

Finding the degrees of freedom for a t-distribution is usually straightforward. It is the number of cases in the sample minus 1. So, if the sample size is 12, the degrees of freedom are 11; if the sample is 21, the degrees of freedom are 20; and so on.

The concept of degrees of freedom refers to the number of observations that are free to vary if the sample mean is to accurately represent the population mean. Remember that in inferential statistics, we are making inferences about a population parameter (such as mean) based on sample statistics. We are really interested in the population mean and would like to represent it, but never know for sure that it has been captured accurately. However, we could make the sample mean equal the population mean with control over just one of the scores.

Suppose that we want to estimate the population mean, and suppose we know that that mean is 8.0. A sample of 6 observations is drawn and includes the following values: 8, 7, 8, 11, 9, 7. The sample mean for the 6 observations is 8.33. This is not the same as 8.0. Yet if we were able to gain control over just one of these scores and change it, we would be able to force the sample mean to be the same as the population mean. To make this happen, let's consider how we would get a mean of 8.0 from 6 observations. Recall that the mean is

$$\bar{X} = \frac{\Sigma X}{N}. \tag{3.1}$$

Since we know the mean and we know N, we can state

$$8 = \frac{\Sigma X}{6}.$$

Multiplying both sides of the equation by 6, we get

$$\Sigma X = 48.$$

So we know that to get a mean of 8, we have to have a sum of 48. Now, let's take just one of our observations from the sample, say, 7. In the sample, the sum was 50. We would like it to be 48, so we take that observation of 7 and reduce it by 2. Now the ΣX is 48 and the mean is 8.

The point of this exercise is to show that if we know the mean, we would always be able to approximate it with a sample by gaining control over just one observation. So all of the observations are free to vary except the one that we tamper with. This is the root of the notion "degrees of freedom." Typically, we do not know the population mean, but we do know that we can achieve it by gaining control of only one observation.

In the practical sense, it is most convenient to remember that the degrees of freedom are $N - 1$. As we advance toward more sophisticated statistical models, the concept of degrees of freedom follows us. In each case, it reflects the number of parameters estimated minus the one that we would need to control to make the sample statistic exactly the same as the population parameter.

Appendix 2 lists the critical values in the t-distribution. This table gives the area in the tails of the t-distribution at a variety of different cutoff points. Let's examine the critical values of t at what is called the ".05 level of significance" for a "two-tailed test." (The terminology is covered in the next chapter.) For now, suffice to say that this is equivalent to the point where 5% of the cases in the sampling distribution lie either above or below the critical value. Earlier we found that this point in the standard normal distribution was 1.96. In other

● BOX 6.9

Why the *t*-distribution Is Sometimes Called Student's *t*

The mathematical derivation of the *t*-distributions was first worked out by an Irish mathematician by the name of W. S. Gossett. At the time, Gossett worked for the Guinness Brewing Company in Dublin. Guinness is well known throughout the world for its ale. The brewery forbid its employees from publishing the results of any research. So when Gossett published his famous article in 1908, he used the pseudonym "Student." Actually, Gossett did not deceive his employer. They knew he was publishing work under the pseudonym and allowed it because it did not reveal any trade secrets. However, since the articles were published under the name "Student," the *t* distribution and the associated *t*-test (see Chapter 7) have become known as Student's *t*.

words, 5% of the cases fell either above a Z-score of 1.96 or below a Z-score of −1.96. Now look at the value in Appendix 2 in the row labeled infinite degrees of freedom under the column for the .05 significance level. You will see that it is 1.96. In other words, with infinite degrees of freedom, the Z- and the *t*-distributions are the same. Now consider the row for 10 degrees of freedom at the .05 α level (middle column in Appendix 2). At this level, the critical value of *t* is 2.228. In other words, to exceed the critical value a *t*-score needs to be larger than a Z-score if there are fewer than an infinite number of degrees of freedom. Since there are 10 degrees of freedom in this case, what is the sample size? Recall that degrees of freedom are $N - 1$, so the sample size is 11. Now look at the top of the column for 1 degree of freedom. Notice here that with only 1 degree of freedom the *t*-value has to be very large (12.706) to achieve the critical level. In other words, at a certain probability level such as .05, the critical value of *t* becomes smaller as the degrees of freedom become larger. Conversely, as the degrees of freedom increase, the critical value of *t* decreases until the *t*- and the Z-distribution are exactly the same with infinite degrees of freedom.

Confidence Intervals in the *t*-Distribution

Finding confidence intervals for the *t*-distribution is very much the same as it was with a standard normal distribution. The only difference is that we now use t-values for certain probability levels instead of Z-values. The formula for the confidence interval using the t-distribution is

$$\bar{X} \pm t_\alpha S_{\bar{X}}. \tag{6.4}$$

Consider the example of the birth weights of babies born in a particular hospital. A sample of 16 is drawn with a mean birth weight of 7.5 pounds and a standard deviation of 1 pound.

The first step is to determine which of the t-distributions we will use to reference our observations. To accomplish this, we must ask how many degrees of freedom we will use. Since there were 16 observations, it will be

$$N - 1 = 15.$$

Let's assume that we are finding the 95% confidence interval, that is, only 5% of the observations will fall outside the interval. Go to Appendix 2 and find the t-distribution for 15 df and the .05 (two-tailed) distribution. This means that 5% of the cases fall above or below this interval. According to the table, the t-value associated with 15 degrees of freedom is 2.131. This will be used as the t_α in the equation. Now, let's find the standard error of the mean for this observation. Remember that the standard error of the mean is

$$S_{\bar{X}} = \frac{S}{\sqrt{N}}$$

$$= \frac{1}{\sqrt{16}} \tag{6.5}$$

$$= \frac{1}{4}$$

$$= .25$$

Next, multiply the t_α-value by the standard error of the mean.

$$t_\alpha S_{\bar{X}} = 2.131 \times .25$$

$$= .53.$$

To find the lower boundary of the confidence interval, subtract this product from the mean:

$$7.5 - .53 = 6.97.$$

Find the upper boundary by adding this product to the mean:

$$7.5 + .53 = 8.03.$$

On the basis of the sample from this hospital, we can say that the probability is 95 in 100 that the mean weight of the babies from the population falls in the interval bounded by 6.97 pounds and 8.03 pounds.

Summary

One of the most important uses of statistics is to make inferences about population parameters on the basis of sample statistics. A population is a set of all possible observations of a specific type. A sample is a subset of observations drawn from the population. The mean and the standard deviation of the sample are unbiased estimates of the population mean and standard deviation if the sample is unbiased and representative of the population. This will occur if the sample is drawn from the population using a random process.

Inferential statistics are used to estimate how well a sample mean (a statistic) represents a population mean (a parameter). To do this, we use the standard error of the mean, which is the sample standard deviation divided by the square root of the sample size. Then a confidence interval around the sample mean is created. The confidence interval is defined by a sample mean plus or minus the Z-score for the probability level of interest times the standard error of the mean. Although we do not know the population mean, we can state the probability that the population mean falls within a defined confidence interval around the sample mean.

Most often in sample statistics we need to use a probability distribution that recognizes that sampling error is greater in small samples. One distribution commonly used is the t-distribution. The t-distribution is actually a family of distributions each with its own degrees of freedom. As the sample size decreases, degrees of freedom also decrease. The fewer the degrees of freedom, the flatter the t-distribution with more area in the tails. Specific applications of the t-distribution are presented in Chapter 7.

In this chapter we have reviewed some of the conceptual material relevant to statistical inference. In the next chapter, we begin applying practical methods of statistical analysis for comparison between groups. The material covered in the last two chapters has been more theoretical than most of the topics covered in this book. This information is important and it may be worth your while to spend a little extra time reviewing it before going on to Chapter 7.

Key Terms

1. **Decision**   A choice from among several alternatives.
2. **Population**   A set of all possible observations of a specific type.
3. **Sample**   A subset of observations drawn from a population.

4. **Random Sample**   A sample in which each member of the population has an equal chance of being selected.

5. **Sampling Distribution**   A theoretical probability distribution of values that could be obtained for some statistic in random samples of a particular sample size from a defined population.

6. **Sampling Error**   The difference between a sample mean and the mean of the sampling distribution. Sampling error is assumed to be random.

7. **Standard Error of the Mean**   The standard deviation in estimating the mean of a population that is based on repeated samples.

8. **Confidence Interval**   An interval for which there is a specified probability of capturing a population mean. For example, for a 95% confidence interval, we can state that the chances are 95 in 100 that the true mean falls within this interval.

9. **Confidence Limit**   The upper or lower boundary of a confidence interval.

10. **Degrees of Freedom**   The number of observations an experimenter would have to control to make a sample mean equal to a population mean. Typically, this is $N - 1$.

Exercises

1.   Write 75 different numbers on pieces of paper. Try to select these numbers to represent some known quantity that you are familiar with. For example, you might use weights of people that you know, shoe sizes, or anything you choose. Place all 75 pieces of paper in a hat. Then draw a random sample of 15 numbers. Calculate the mean for the sample. Put the 15 numbers back in the hat, shake the hat, and draw a second sample of 15 numbers. Now calculate the second mean. Repeat this five different times. Are the sample means equal to one another?

2.   Does a large sample size ensure that the sample mean will be equal to the population mean?

3.   What is the difference between a parameter and a statistic?

4.   Suppose you have drawn a sample of 100 observations and calculated the variance as 25. What is the standard error of the mean?

5.   You have drawn a sample and found the standard deviation to be 32.7. There were 74 cases in this sample. What is the standard error of the mean?

6.   You have observed a sample mean of 62.4. The standard error of the mean is 4.21. What is the 95% confidence interval?

7. A sample of 42 observations has yielded a sample mean of 16.4 and a standard deviation of 3.21. What are the limits of the 95% confidence interval?

8. A sample has a mean of 12.2 and a standard deviation of 6.1. There were 14 observations. How many degrees of freedom would there be using the t-distribution? Find the 95% confidence interval based on the t-distribution.

7

Methods for Comparing Two Sets of Observations

In experimental sciences comparisons between groups are very common. Usually, one of the groups is treated and is called the *experimental* group, and the other group is untreated and is called the *control* group. After the treatment has been administered scientists often want to determine whether the two groups differ. Yet, as we have seen in the preceding chapters, groups will differ by chance. The task for the scientists is to determine whether any observed differences between the groups should be attributed to chance or to the experimental intervention. One statistical tool that helps the scientists make this decision is called the *t*-test. It is one of the most commonly used statistical tests in science and one that you will come across repeatedly as you read scientific journals.

Actually, there is not just one *t*-test but rather a family of statistical tests that all use the *t*-distribution. Some of these tests are comparisons between a sample mean and a hypothetical population mean, comparisons between two sets of observations for one group of individuals, and comparisons between observations made for two independent groups. The procedures to perform all of these tests, and more, are covered in this chapter. Before introducing the procedures, however, it may be worthwhile to review the research situations that require the different types of *t*-tests.

■ Examples of *t*-Test Research Problems

1. *Comparison of a Sample Mean to a Hypothetical Population Mean.* An educational researcher wants to know whether achievement scores in a particular

school district are lower than the average score for the state. The mean score for the state as a whole is published. He takes a random sample of students from the particular school district, calculates the mean, and now determines whether they differ significantly from the state averages.

2. *Comparison Between Two Scores in the Same Group of Individuals.*   A clinical psychologist wants to learn whether biofeedback reduces the incidence of headaches. He takes one group of patients with a consistent history of headaches. He measures the number of headaches they get per week for a one-month period and takes the average for each patient. All patients then receive biofeedback and the psychologist measures the average number of headaches they get in the first month after the biofeedback training. He then compares the average number of headaches before treatment with the average number after treatment.

3. *Comparison Between Observations Made on Two Independent Groups.*   An investigator wants to determine whether systematic desensitization helps reduce anxiety among college students in public speaking courses. Students in a public speaking course are randomly assigned either to a systematic desensitization condition or to a control group in which they receive an equal amount of attention but do not receive the desensitization. After these treatment and control procedures have been administered, judges (who are not aware of the particular condition each student was assigned to) attend the speech classes and rate students' anxiety while they are giving their presentations. The researcher needs to determine whether the observed differences in anxiety levels between the two independent groups are statistically significant.

Research Concepts

The study of statistics can help you understand and perform research. To understand research, we have to review some of the terms that appear frequently in research reports. Experiments determine the causes of behavior, illness, or other outcomes. This is a major goal of behavioral and health sciences. In behavioral science, we are constantly asking the question "Why?" Why do people like one another? Why are we sometimes aggressive? Under what psychological or medical circumstances does a treatment help someone? You might think of experiments as attempts to answer questions about behavior in a formalized and highly controlled manner. Nonexperimental research methods share many features with experiments, but they have a lesser degree of control.

Experiments involve both constants and variables. Many influences are the same (constant) for all the participants in the study. For example, the room in which the experiment is conducted should be the same as should the instruments used to make observations. The experimenter usually manipulates one or more *variables* and attempts to determine what effects this tinkering produced. A variable may take on different values for different participants in the study. The variable manipulated by the experimenter is known as the *independent variable*. Thus, in experimental studies, it is appropriate to think of the independent variable as the manipulation. Of course, to determine what effect the manipulation has, we must measure another variable that we believe may be affected by variation of the independent variable. This is known as the *dependent variable*. Try to remember that the independent variable is the manipulation and the dependent variable is the outcome measure.

Sometimes the manipulation of the independent variable involves the administration of experimental *treatments*. Subjects are given different treatments, and the different treatments are called levels of the independent variable. In a typical experiment we may give the treatment to one group and withhold it from another group. In this case, the independent variable has two treatment levels. The group that received the treatment is called the **experimental group,** and the group that did not receive the treatment is called the **control group.** Some experiments have several levels of the experimental treatment and may have several experimental groups.

In true experiments, the experimental and control groups differ only for random reasons before the administration of the experimental treatments. If, after the administration of the treatment, the experimental and control groups differ, then the differences can be attributed to the treatment. It is possible to make this causal statement because *extraneous* variables have been controlled. An extraneous variable is one that might affect the relationship between an independent and a dependent variable or may affect the dependent variable independently from the treatment. The purpose of control in experimental studies is to hold constant the influence of extraneous variables.

Let us take the example of a research study on the effects of television violence on aggressive behavior among children. Correlational evidence suggests that those children who watch violent adult role models on television tend to be more aggressive than children who watch less TV violence. These studies, however, do not show that television causes aggression. It may be that exposure to television violence causes aggression or that aggressive children prefer to watch violent television shows. To determine the direction of causation, it is necessary to control extraneous variables as well as the exposure to television. (Correlational techniques are presented in Chapter 10.)

In an experimental setting, we might begin by dividing children into two groups. This has to be done in a way that does not produce systematic differences between the groups. By randomly assigning the children to the different conditions, the experimenter eliminates *systematic* biases or differences between groups prior to the experiment—that is, the initially aggressive children have an equal chance of being assigned to the experimental group or the control group. The control group is not shown television violence but might be exposed to equally exciting nonviolent television. The control group is necessary to determine the level of aggressive behavior that occurs as a function of viewing nonviolent television. The experimental group is exposed to television violence. Later, both groups of children are observed on the playground, and the number of their physical and verbal aggressive behaviors are recorded by trained observers.

In this experiment, television exposure is the independent variable. The violence-viewing and nonviolence-viewing groups are the experimental and control groups and define the levels of the independent variable. The measures of physical and verbal aggression on the playground are the dependent variables. Extraneous variables, such as the amount of violence the children usually watch, are controlled because each child has an equal chance of being assigned to either the experimental or the control group. Because the children are assigned to the two experimental groups by a random process, the groups differ only by chance before the experimental treatment.

After the treatment the groups are compared. Statistical methods then determine the chances that any observed differences between the groups would have occurred by chance alone. If we are confident that the differences were not the result of chance, we conclude that it was the exposure to television that caused the two groups to differ in aggression.

It is important to emphasize that true experiments require that subjects be assigned to experimental and control groups by some random process. If the assignment is not random, then the statistical methods cannot adequately specify the chances that the groups differ prior to the experimental treatments. This should become more obvious as we look at some of the variations on the experimental method.

Some Basic Rules of Statistical Testing

Throughout the remainder of this book we present a variety of procedures to evaluate statistical significance. Although the procedures differ from one another and are used in different situations, there are some common elements. All of these procedures are based on the ratio of observed to expected differences. The top half of the ratio is the observation. For example, in the t-test it is usually the observed differences between means. The bottom half of the ratio is the standard error for that same observation. This is an estimate of the extent to which the means would be expected to differ by chance alone. Recall from Chapter 4 that the standard deviation is an estimate of the average expected deviation around the mean. The standard error of the mean (Equation 6.1) is the standard deviation of the sampling distribution. The standard error of the mean is an approximation of the average deviation we would expect in estimating the population mean from a sample mean. In the t-test we look at the difference between means, so the bottom half of t-equation is the standard error of the difference between means. In other words, we examine the observed differences between means divided by an approximation of the average deviation we would expect due to sampling error. If these ratios are relatively large, then the top half—the observed deviation—is greater than what would be expected by chance. If it is less than 1, then the amount of variation expected by chance is greater than the amount observed, and we should not conclude that any observed differences are significant. As we go through the various tests, keep this principle of the ratio of observed to expected scores in mind. It will come up repeatedly.

Statistical Tests for the Mean of a Single Population

The first statistical significance test we consider is for a single population mean. Here are some examples.

1. We want to determine whether a particular baseball team (for example the San Diego Padres) is "a good hitting" team. You know the mean batting average for the entire league, so you take the team batting average for the Padres to see if it is significantly higher or lower than the population mean.
2. You want to determine whether the mean achievement in your child's school is lower than the state average. Test scores for your child's school and the state as a whole are available.
3. A doctor draws a sample of blood from a 19-year-old college student and analyzes it to determine the number of white cells. She wants to determine whether the student has an abnormally high white blood count.

Each of these problems could be analyzed using a one-sample Z- or t-test. The choice between the Z- and t-test hinges on whether the population standard deviation is known or unknown. In the first example with the batting averages for various baseball teams, the population standard deviation is probably known. During the baseball season the batting averages for all players are often published in the Sunday paper. In the second example involving analysis of performance of a particular school district, the population standard deviation may or may not be known. In other words, you may or may not be able to get a standard deviation of test scores for the entire state because the state data may be based on selected samples of students. When you are uncertain, it is most conservative to assume that the population standard deviation is unknown. In the third example of white blood cell count, the population standard deviation is unknown. It would not be possible to take an infinite number of samples of blood to determine the population value.

The choice of a statistical test for evaluation of a single population is based on the following rules.

- If the population standard deviation is known, use the Z-distribution.
- If the population standard deviation is unknown, use the t-distribution.

In other words, your choice depends on whether information about the population standard deviation is known or derived from an estimate. The formulas for the Z-test and t-test are very similar.

■ The Z-Test

The formula for a Z-test is very much the same as the formula for the Z-score covered in Chapter 4. Remember that a Z-score was

$$Z = \frac{X - \bar{X}}{S} \tag{4.2}$$

The formula for a Z-test for evaluating a single population is

$$Z = \frac{\bar{X} - \mu}{\sigma_{\bar{X}}}, \tag{7.1}$$

where $\bar{X}$ is the sample mean, μ is the hypothesized value of the population mean, and $\sigma_{\bar{X}}$ is the standard error of the mean when the population standard deviation is known. The standard error of the mean is

$$\sigma_{\bar{X}} = \frac{\sigma}{\sqrt{N}}. \tag{6.1}$$

Statistical tests usually involve the expression of some observed difference over some estimate of the average variation in that difference. For example, a Z-score states the difference between a score and the mean divided by an approximation of the average variation of a score around the mean. In the case of the Z-test we are expressing the difference between an observed sample mean and a hypothetical population mean. This is the top part of the equation. The bottom half of the equation represents an estimate of the average error in estimating the population mean from a sample. This is the standard error of the mean.

Now let's calculate the Z-test for some examples. The first example comes from professional sports. The 1982–83 National Football League season was shortened due to a strike. That year, only 9 regular season games were played. In most seasons playoff teams are selected to represent different divisions. Yet because of the strike, teams were selected on the basis of the best record, and this left some divisions poorly represented. One of these divisions was the National Football Conference (NFC) Western Division. Some sports commentators argued that this was good because the teams in the NFC West were of lower quality. Our question is whether the teams the NFC West indeed had a significantly poorer record than all teams that year.

The mean number of games won by all 28 teams in the National Football League that year was 4.46. The standard deviation was 1.88.* The teams in the NFC West and the number of games they won are as follows.

Atlanta Falcons	5
New Orleans Saints	4
San Francisco 49ers	3
Los Angeles Rams	2

The mean of the sample is

$$\bar{X} = \frac{\Sigma X}{N} = \frac{5 + 4 + 3 + 2}{4} = 3.5.$$

Now we are ready to calculate the value for the statistical test. Since the population standard deviation is known, we will use the Z-test. The formula is

$$Z = \frac{\bar{X} - \mu}{\sigma_{\bar{X}}}.$$

Remember that

$$\sigma_{\bar{X}} = \frac{\sigma}{\sqrt{N}},$$

so

$$\sigma_{\bar{X}} = \frac{1.88}{\sqrt{4}}$$

$$= .94.$$

Now we can calculate the Z-score.

$$Z = \frac{\bar{X} - \mu}{\sigma_{\bar{X}}}$$

$$= \frac{3.5 - 4.46}{.94}$$

$$= -1.02$$

So the Z-value is −1.02. Before we interpret the meaning of this Z-value let's consider the same example as though the mean and the standard deviation of the population were unknown. In this case, we use the t-test rather than the Z-test since we assume that the population standard deviation is unknown and

---

*In a 9-game season we would expect this to be 4.5 because there is a winner and loser in each game. Thus, half of all games should be won and half should be lost. However, in the 1982–83 season there was one tie. Thus, the mean was 4.46 rather than 4.5.

that we must estimate it from the sample standard deviation. The sample standard deviation for the NFC West teams was 1.29.* Thus, the standard error of the mean becomes

$$S_{\bar{x}} = \frac{S}{\sqrt{N}} = \frac{1.29}{\sqrt{4}} = .64.$$

The t-value is

$$t = \frac{\bar{X} - \mu}{S_{\bar{x}}}. \tag{7.2}$$

Note that the only difference between this t-formula and the Z-formula above is that we use $S_{\bar{x}}$ instead of $\sigma_{\bar{x}}$. Let's assume that we do not know the population mean. Yet theoretically it should be 4.5 because half of the games should be won and half should be lost. Thus, in the 9-game season, the expected average level of performance should be 4.5. The t-value becomes

$$t = \frac{3.5 - 4.5}{.64} = \frac{-1}{.64} = -1.56.$$

To summarize, we have obtained a Z-value of -1.02 under the assumption that the population standard deviation is known, and a t-value of -1.56 under the assumption that the population standard deviation is unknown. Typically, you would not calculate both of these values. Instead, you would select one approach based on whether the population standard deviation was available to you.

Interpretation of Statistical Tests

Now that we have calculated the Z- and t-values, what do they mean? In Chapter 6 we reviewed two characteristics of statistical inference that are similar to the principles of criminal law. First, there was the assumption that means do not differ. This is the null hypothesis. This is analogous to the assumption in criminal law known as presumption of innocence. We presume that a defendant is innocent until proven guilty. Second, there is the notion that if we decide that means differ, it must be on the basis of substantial evidence. This is analogous to the legal principle of proof beyond a reasonable doubt. Stating the null hypothesis and deciding upon evidence used to reject the null hypothesis are very important principles in statistical inference. Therefore, we will review them in some detail.

---

*The standard deviation was obtained for the number of wins (5, 4, 3, 2).

■ The Null Hypothesis

In scientific inquiry we often advance hypotheses, or hunches. Science by nature is full of skepticism. The scientist never wants to believe anything unless there is proof. A scientist might not believe someone who says that a new drug works, a new therapy is effective, or a new teaching method is stimulating. The scientist's assumption is that these approaches do not work and that the burden of proof is on the person making these claims to provide evidence for their value.

In formal studies and experiments, scientists typically begin with the assumption that there will be no differences between different groups. In the case of testing a single sample mean against a hypothetical population, the assumption is that the sample mean does not differ significantly from the population parameter. In other words, it is assumed that the sample can be considered representative of the population. As we have seen several times, different sample means differ from one another. It would actually be rare to draw two samples and obtain exactly the same mean. Under the null hypothesis, we assume that these observed differences are due to sampling error. However, there may be circumstances in which the null hypothesis becomes untenable. For example, there may be substantial evidence that the null hypothesis is incorrect. Under these circumstances, we would reject the null hypothesis in favor of an alternative hypothesis.

■ Alternative Hypothesis

In statistical analysis, the null hypothesis is typically tested against an alternative hypothesis. The alternative hypothesis usually states that observed differences are not attributable to sampling error.

To summarize, the research investigator has two hypotheses to choose between. The null hypothesis states that observed differences are attributable to chance, whereas the alternative hypothesis states that observed differences are not attributable to chance. Selecting between these two hypotheses requires formal decision rules.

■ Decision Rule

Choosing between the null and alternative hypotheses would be difficult if the scientist did not have a well-defined plan. To make this decision the scientist develops a specific plan of action before the experiment begins. This plan or decision rule sets forth the circumstances under which the null hypothesis will be rejected.

The first step in the execution of this plan is to provide a precise statement of the null hypothesis. Usually we refer to the null hypothesis as H_0. In the case of the single sample test we might state that the sample mean equals the population mean, or

$$H_0: \quad \overline{X} = \mu, \quad \text{or} \quad \overline{X} - \mu = 0. \tag{7.3}$$

This would be tested against the alternative hypothesis that the sample mean does not equal the population mean. The alternative hypothesis is usually called H_1. We would state the alternative hypothesis as

$$H_1: \quad \overline{X} \neq \mu. \tag{7.4}$$

In repeated random samples from the same population, we will obtain a sampling distribution of sample means. That sampling distribution of sample means might look like the one shown in Fig. 7.1.

Figure 7.1 shows the error in estimating a population mean over a theoretically large number of samples. It is most probable that the sample mean will be close to the population mean as reflected by the high center portion of the figure. The greater the discrepancy between the sample and population mean, the lesser the probability that the difference will occur by chance. In the decision rule, we say that we will assume that the null hypothesis is correct until we prove otherwise. Specifically, we say that differences between the observed mean and the hypothesized population mean must be very improbable by chance alone before we reject the null hypothesis and accept the alternative hypothesis.

■ Significance Levels

How improbable must this difference be before we reject the null hypothesis and accept the alternative hypothesis? Although there are no hard and fast

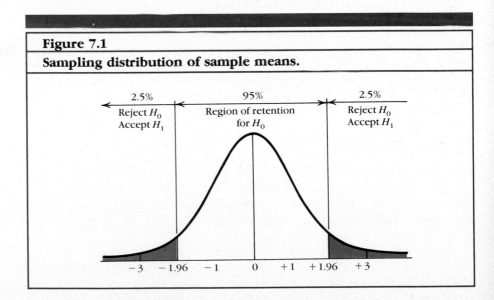

Figure 7.1

Sampling distribution of sample means.

rules, it is common in statistics to make decision rules based on specific cut-off points, such as .05 and .01. The .05 cutoff point or *significance level* means that the probability of an observed difference must be less than .05 by chance alone before we reject the null hypothesis and accept the alternative hypothesis. In other words, the probability of a difference by chance alone must be relatively rare (less than 5 in 100). In the case of the .01 level, the probability of observing a difference of this magnitude by chance alone must be less than 1 in 100.

If we repeatedly drew samples from the same population and took the difference between the sample and population means, we could create a sampling distribution that looks like the one in Fig. 7.1. Then we could create a 95% confidence level interval around the mean of the distribution. In practice, however, we typically have just one sample mean. In our National Football League example, we have only one sample mean for the NFC West teams. We want to see if this sample mean differs from the population mean. To do this we subtract the population mean from the sample mean and divide it by the standard error of the mean using the formula for the *t*-value:

$$t = \frac{\bar{X}_1 - \mu}{S_{\bar{X}}}. \tag{7.2}$$

Now we want to determine whether or not the difference we observe is unusual. If the difference we observe falls within the 95% confidence interval around the mean of the sampling distribution, we will fail to reject the null hypothesis. In Fig. 7.1 this area is labeled the region of nonrejection, or H_0. Notice that we do not formally accept H_0, we simply *do not reject it.* The rationale for not formally accepting the null hypothesis is given in Box 7.1. If the difference falls in the top 2.5% or the bottom 2.5% of the sampling distribution, we will reject H_0.

To evaluate the difference between the mean performance of the NFC West teams against the population mean, we use a formal statistical test. We begin by stating a decision rule. First we consider the Z-distribution. Figure 7.2 shows the hypothetical sampling distribution around the sample mean. Using the Z-distribution we set cutoff points for levels below and above the mean; we use Appendix 1 to determine which Z-values are associated with the top and the bottom 2.5% of the distribution. According to the table these values are positive 1.96, and negative 1.96. If you do not understand how we got these values, go back and review the material in Chapter 4.

With this information, we are ready to state the null hypothesis. The null hypothesis is

H_0: $\bar{X} = \mu.$

This is tested against the alternative hypothesis.

H_1: $\bar{X} \neq \mu$

The decision rule is stated as follows. If the observed Z is greater than -1.96 and less than $+1.96$, do not reject the null hypothesis. If the observed Z

Retaining the Null Hypothesis

Technically, we never *accept* the null hypothesis. Actually, experimental results tell us more when we reject the null hypothesis then when we fail to reject it. If we do not reject the null hypothesis we are saying that we do not have sufficient evidence to say that the observed differences between the means are significant according to our strict criteria. In the criminal justice system, evidence against the accused must be convincing beyond a reasonable doubt for there to be a conviction. If a person is judged not guilty this does not mean they are innocent. It only means that the evidence left at least a reasonable doubt. In statistical hypothesis testing we also have strict criteria for rejection of the null hypothesis. If we do not reject the null hypothesis we are saying that either the null or the alternative hypothesis could be true. However, we do not reject the null hypothesis unless the probability is small that the observed differences occurred by chance alone.

Thus, instead of saying that we "accept" the null hypothesis, we typically say that we "fail to reject" or "retain" it.

is less than -1.96 and greater than $+1.96$, reject the null hypothesis and accept the alternative hypothesis.

The observed Z-value for the National Football Conference (NFC) example was -1.01. This clearly falls within the range of retaining the null hypothesis. Figure 7.2 shows this particular Z-value within the region of acceptance of the null hypothesis. The conclusion, on the basis of our test, is to retain the null hypothesis. Retention of the null hypothesis means that differences between the observed mean and the population mean should be attributed to sampling error. It is very important to realize that "acceptance" (technically, retention) of the null hypothesis does *not* mean accepting that the sample mean and the population mean are the same. It only means that observed differences should be attributed to chance. Retention of the null hypothesis is actually an endorsement for the statement: "There is not enough evidence to attribute observed differences to anything systematic." This is similar to the legal pronouncement that someone is not guilty. Not guilty does not necessarily mean innocent. It only means that there was not evidence beyond a reasonable doubt for guilt. The system is designed to have strict requirements for the assignment of guilt and penalty. Similarly, scientific systems are designed to err on the side of conservatism.

The procedure for evaluating the t-test is very much the same. The only difference is that we use a t-distribution rather than a Z-distribution. For our NFC example we determined the point on the distribution for rejection of the null hypothesis from Appendix 1. This time you can consider the NFC example using the t distribution. Remember that determining the appropriate cutoffs re-

Figure 7.2

Sample distribution of sample means for Z-test and NFC example.

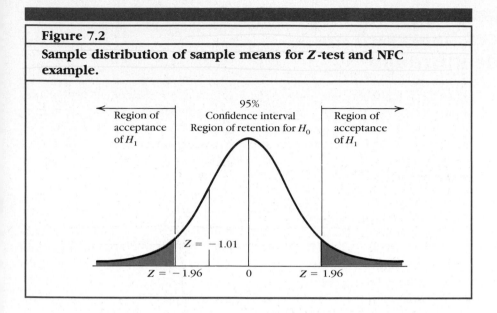

quires that you enter the table for specific degrees of freedom. In the NFC example, there were 4 teams in the sample, so the degrees of freedom are

$$\text{df} = N - 1$$
$$= 4 - 1 = 3.$$

Now look at Appendix 2. The task in examining this table is to find the cutoff value for the t-distribution at the .05 level with 3 degrees of freedom. So find the row for 3 degrees of freedom and go across the table to the column labeled .05 for a two-tailed test. The value in the table is 3.182. Using this information we can construct the decision rule. The null hypothesis is

$$H_0: \quad \overline{X} = \mu,$$

which is evaluated against the alternative hypothesis

$$H_1: \quad \overline{X} \neq \mu.$$

The decision rule states that if the observed t is greater than -3.182 or less than $+3.182$ we retain H_0. If the observed t is less than -3.182 or greater than $+3.182$ we reject H_0 and accept H_1. In the NFC example the observed t was -1.56; this is clearly within the region of retention of the null hypothesis. Thus, we conclude that the differences between the average performance of NFC teams and the hypothetical expected level of performance for these teams is attributable to sampling error. In other words, the differences are not statistically significant.

Differences Between Two Independent Groups

Procedures for testing for differences between two independent groups are similar to the ones we have just discussed for comparing a sample mean to a population mean. Actually, the comparison of two independent groups is among the most common tasks in scientific investigations. These comparisons are typically done using the *t*-test for independent samples. Here are some examples of the use of *t*-tests for independent samples.

1. A psychologist wants to determine whether behavior modification is effective for reducing the number of "acting out" behaviors among elementary school children. He randomly chooses 50 classrooms for the program and uses an additional 50 classrooms for control. Then the psychologist monitors the number of acting out behaviors over three months and compares the classrooms exposed to the program with the ones not experiencing it.
2. A school administrator wants to determine whether more students smoke in schools that have liberal smoking policies. She believes that schools with a permissive smoking policy encourage teenage smoking. She draws a random sample of schools that allow smoking and tabulates the percentage of students who report on a questionnaire that they smoke. Then she does the same for a random sample of schools that do not permit smoking. A *t*-test determines whether the mean number of smokers is different in the two samples.
3. A developmental psychologist wants to know whether infant boys and infant girls differ in their response to auditory stimulations seven hours after birth. She draws a random sample of babies from each of these two groups, measures the response, and compares the results for the two genders.

The difference between the independent group tests and those discussed in the last section is that two independent samples are drawn. The rationale for these tests is that two sample means drawn from the same population should differ only by sampling error. A large difference between these two means may imply that they were not in fact drawn from the same population or from populations with the same mean. Observed differences might mean, for instance, that boy and girl neonates actually differ in response to auditory stimulation. In other words, the population mean for boys and the population mean for girls may actually be different. Similarly, a difference between the number of smokers at schools that permit or do not permit smoking may imply that these two types of schools are not random samples from the same population at least in relation to the number of students who smoke. Earlier in the chapter, it was noted that most statistical tests follow the same logic. The top half of the equation is the observed difference, and the bottom half is the average difference that might be expected by chance. In the case of the independent group's *t*-test, the bottom portion of the equation is the standard error of the difference between means.

■ Standard Error of the Difference Between Means

The task in the independent groups *t*-test is to evaluate the differences between sample means. This is done by utilizing a sampling distribution of differences between the sample means. In other words, it would be possible to repeatedly draw samples for two different groups and take their means. Using the example of response to auditory stimulation among neonate boys and girls, we could go to the hospital and draw a sample of 10 boys and a sample of 10 girls. Then we would calculate the mean for the boys and the mean for the girls. This experiment could be repeated over and over and over again. Each time we drew the sample of boys and girls, we could find the difference between the boy and girl means. Since we are doing this over and over again, we can eventually begin to create a sampling distribution of the differences between sample means. This sampling distribution will have its own mean and its own standard deviation. The standard deviation of the sampling distribution would be the standard error of the differences between sample means. This is a difficult concept and worth thinking about. Remember that in this distribution, each entry is the observation of the difference between a pair of sample means. That is why it is called the sampling distribution of differences between sample means. To calculate the *t*-test, we evaluate the difference between the means for two different groups relative to standard error of differences between means. The formula for this *t*-test is

$$t = \frac{\overline{X}_1 - \overline{X}_2}{S_{\overline{X}_1 - \overline{X}_2}}, \tag{7.5}$$

where $\overline{X}_1$ is the sample mean for group 1, $\overline{X}_2$ is the sample mean for group 2 and $S_{\overline{X}_1 - \overline{X}_2}$ is the standard error of difference between sample means.

The crucial portion of this is the bottom half of the equation which is the standard error of the differences between sample means. This is usually called the *standard error of the difference*. The formula for the standard error of the difference is somewhat complex looking (see Box 7.2). Actually, it is very similar to the notion of the standard error of the mean that we introduced in Chapter 6. However, it is necessarily more complicated because we are dealing with two different samples that may have two different standard deviations. Some textbooks give simplified versions of these formulas that can be used in special cases. But since most students nowadays have access to calculators or computers that do *t*-tests, it may be most appropriate to use this general formula. Box 7.2 gives the step-by-step breakdown for calculating the independent groups *t*-test.

● BOX 7.2

Step-by-Step Calculation of the *t*-Test

Question: Did the San Diego Chargers score significantly more points than the New England Patriots during the 1982–83 season?

Formula:

$$t = \frac{\bar{X}_1 - \bar{X}_2}{\sqrt{\left[\frac{(N_1 - 1)S_1^2 + (N_2 - 1)S_2^2}{N_1 + N_2 - 2}\right] \cdot \left[\frac{1}{N_1} + \frac{1}{N_2}\right]}}$$

Data: Points scored by the San Diego Chargers (Team 1) and the New England Patriots (Team 2) during 9 games in the 1982–83 season.

TEAM 1 (CHARGERS)		TEAM 2 (PATRIOTS)	
Opponent	Points	Opponent	Points
Broncos	23	Colts	24
Chiefs	12	Jets	7
Raiders	24	Browns	7
Broncos	30	Oilers	29
Browns	30	Bears	13
49ers	41	Dolphins	3
Bengals	50	Seahawks	16
Colts	44	Steelers	14
Raiders	34	Bills	30

STEPS

1. Calculate the mean for Team 1, symbolized as $\bar{X}_1$.

$$\bar{X}_1 = \frac{\Sigma X}{N} = \frac{(23 + 12 + 24 + 30 + 30 + 41 + 50 + 44 + 34)}{9}$$

$$= \frac{288}{9}$$

$$= 32.00$$

2. Calculate the variance for Team 1, symbolized as S_1^2.

$$S_1^2 = \frac{\Sigma X_1^2 - \frac{(\Sigma X_1)^2}{N_1}}{N_1 - 1} = \frac{10322 - \frac{(288)^2}{9}}{9 - 1}$$

$$= \frac{10322 - 9216}{8}$$

$$= 138.25$$

(If you have trouble with this step, review Box 3.8.)

3. Calculate the mean for Team 2, symbolized as $\bar{X}_2$.

$$\bar{X}_2 = \frac{\Sigma \bar{X}}{N_2}$$

$$= \frac{(24 + 7 + 7 + 29 + 13 + 3 + 16 + 14 + 30)}{9}$$

$$= \frac{143}{9}$$

$$= 15.89$$

4. Find the variance for Team 2, symbolized as S_2^2.

$$S_2^2 = \frac{\Sigma X_2^2 - \dfrac{(\Sigma X_2)^2}{N}}{N - 1}$$

$$= \frac{3045 - \dfrac{(143)^2}{9}}{9 - 1}$$

$$= 96.61$$

5. Find $(N_1 - 1)S_1^2$ by multiplying the results of Step 2 by $N_1 - 1$.

$$(N_1 - 1)S_1^2 = (9 - 1)138.25$$
$$= 8 \times 138.25$$
$$= 1106.00$$

6. Find $(N_2 - 1)S_2^2$ by multiplying the results of Step 4 by $N_2 - 1$.

$$(N_2 - 1)S_2^2 = (9 - 1)96.61$$
$$= 8 \times 96.61$$
$$= 772.88$$

7. Find

$$N_1 + N_2 - 2 = 9 + 9 - 2$$
$$= 16.$$

8. Obtain $\left[\dfrac{(N_1 - 1)S_1^2 + (N_2 - 1)S_2^2}{N_1 + N_2 - 2}\right]$ using

the results of Steps 5 through 7.

$$\left[\frac{1106.00 + 772.83}{16}\right] = \frac{1878.88}{16}$$
$$= 117.43$$

9. Find $\dfrac{1}{N_1} = \dfrac{1}{9} = .11$.

10. Find $\dfrac{1}{N_2} = \dfrac{1}{9} = .11$.

11. Obtain $\left[\dfrac{1}{N_1} + \dfrac{1}{N_2}\right]$ by summing the results of Steps 9 and 10.

$$.11 + .11 = .22$$

12. Get the product of Steps 8 and 11.

$$(117.43)(.22) = 25.83$$

13. Take the square root of Step 12.

$$\sqrt{25.83} = 5.08$$

This is the bottom half of the formula or

$$\sqrt{\left[\frac{(N_1 - 1)S_1^2 + (N_2 - 1)S_2^2}{N_1 + N_2 - 2}\right] \cdot \left[\frac{1}{N_1} + \frac{1}{N_2}\right]}$$

14. Find $\bar{X}_1 - \bar{X}_2$ by subtracting Step 3 from Step 1.

$$\bar{X}_1 - \bar{X}_2 = 32.00 - 15.89 = 16.11$$

This result forms the top half or numerator of the t-ratio.

15. Obtain the t-ratio by dividing the results of Step 14 by Step 13.

$$t = \frac{16.11}{5.08} = 3.17$$

16. Determine df as

$$N_1 + N_2 - 2 = 9 + 9 - 2$$
$$= 16.$$

17. Evaluate the null hypothesis (see text).

In this example, we attempt to determine whether the San Diego Chargers averaged more points per game than the New England Patriots during the 1982–83 National Football League season. The box emphasizes step-by-step calculations used to obtain the t-value. To apply the decision rule, we first state the null hypothesis and the alternative hypothesis. In this case, the null hypothesis would be

H_0: $\mu_1 = \mu_2$.

In other words, it would state that the population mean for points scored by the San Diego Chargers is equal to the population mean of points scored by the New England Patriots. For each team, we estimate the population mean based on a sample of its performance. In the 1982–83 football season there were 9 such samples for each team. These were the 9 games played during that year (this was the year of the NFL players' strike and the season was shorter than usual).

The null hypothesis is tested against the alternative hypothesis, which might be stated as:

H_1: $\mu_1 \neq \mu_2$.

Since the 1982–83 football season was a shortened one and one in which there was considerable turmoil among the players, we want to have a stringent criterion for rejecting the null hypothesis. So we pick the .01 alpha or significance level. This is the probability of rejecting the null hypothesis by chance. In other words, by selecting the .01 level, we are saying that we would only reject the null hypothesis by chance in one out of every 100 independent trials. Appendix 2 gives the critical values of the t-distribution. In this example, there are 9 scores for each team so the degrees of freedom are

$(N_1 - 1) + (N_2 - 1)$,   or
$N_1 + N_2 - 2 = 9 + 9 - 2 = 16$.

Check the required t-value to reject the null hypthesis at the .01 level with 16 degrees of freedom. This is done by going down the column labeled *df* at the left side of the table until you reach 16. Then go across the table for the row labeled 16 to the column labeled .01 for the two-tailed test. The value there is 2.921. Remember, this is a two-tailed test, so the decision rule can be stated as follows.

If the observed t is greater than or equal to -2.921 and less than or equal to $+2.921$, retain H_0.

If the observed t is less than -2.921 or greater than $+2.921$, reject H_0 and accept H_1.

The example shown in Box 7.2 reveals that the observed t is 3.17. This exceeds $+2.921$. Therefore, the null hypothesis is rejected in favor of the alternative hypothesis. The conclusion is that in repeated random sampling of size 9 from a single population, the probability of obtaining by chance differences equal to or larger than those observed is less than 1 in 100. Thus, the observed differences are considered improbable by chance alone.

With regard to this specific example, the analysis suggests that the San Diego Chargers scored significantly more points during the 1982–83 season than did the New England Patriots.

One- and Two-Tailed Tests

So far we have talked only about two-tailed tests. These are also called *nondirectional* tests. In a nondirectional test a significant difference can be either above the hypothetical population mean or below it. There are a number of occasions in which we are only interested in differences in one direction. For example, if we think that a drug may have a harmful side effect we might be interested only in differences at one end of the distribution. Let's consider an example of a new drug designed to improve memory. Although the drug seems effective for this purpose, it also seems to produce headaches. A random sample of volunteers is divided into two groups. One group receives the new drug and the other a placebo. (A placebo is a treatment with no known effects.) An investigator wants to know if those given the drug have a significantly higher level of headache. She states that the observation that the drug reduces headaches in relation to the control condition is very unlikely and not of particular interest at the present time.

The null hypothesis is stated to reflect that the investigator is only interested in differences in one direction. It is

H_0: $\mu_1 \le \mu_2$,

which is tested against the alternative directional hypothesis

H_1: $\mu_1 > \mu_2$,

where μ_1 is the population mean of the treatment group and μ_2 is the population mean of the placebo group.

There were 15 subjects in each of the two groups in the experiment, so the degrees of freedom for the t-test are

$15 + 15 - 2 = 28.$

Figure 7.3 shows the sampling distribution with the region of acceptance of H_1 all in upper portion of the distribution. This looks different from Fig. 7.2, which had the region of acceptance at H_1 divided between the upper 2.5% and the lower 2.5% of the distribution. The t-value associated with the upper 5% of the distribution can be found in Appendix 2.

Since there are 28 degrees of freedom, we find the row for 28 under *df.*

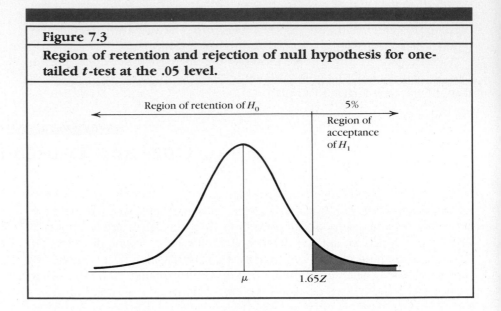

Figure 7.3

Region of retention and rejection of null hypothesis for one-tailed t-test at the .05 level.

Next, we locate the column labeled .05 *for the one-tailed test*. This is exactly the same column that is labeled for the .10 level of significance for the two-tailed test. It is the same column because it is the point that marks the upper 5% of the distribution. In the two-tailed test at the .10 level of significance, we are considering the top 5% and the bottom 5% of the distribution. Drawing pictures of distributions showing areas at the top and bottom may help you understand this concept. An example is shown in Fig. 7.4.

To set the cutoff level we go across the row for 28 degrees of freedom until we hit the column for the .05 one-tailed level. The value listed there is 1.701. We can state the decision rule as follows.

If the observed $t \leq 1.701$, retain H_0.

If the observed $t > 1.701$, reject H_0 and accept H_1.

Now suppose that the experimenter investigating the side effects of the memory drug actually found more headaches among those who took the drug. The t-value for this observation was 1.73. What would her decision be?

Since $1.73 > 1.701$, her decision would be to reject the null hypothesis and accept the alternative hypothesis. The interpretation of this would be that the greater number of headaches observed among those taking the drug is not attributable to sampling error. In other words, she might be inclined to conclude that the drug caused headaches.

Now let's suppose that the investigator had studied a different drug. This time, let's assume that she observed fewer headaches among those who had

Figure 7.4

Specific example for *t*-distribution and 28 degrees of freedom.
The top portion shows the region of rejection (shaded area) for a two-tailed test. In this case, it is a *t*-score ≥ 2.05 or ≤ -2.05. The bottom portion shows the region of rejection for a one-tailed test in the upper part of the distribution (shaded area, $t \geq 1.70$).

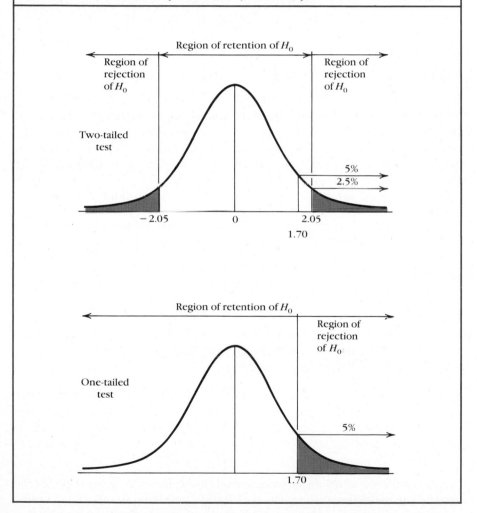

taken the experimental drug. Actually, there were many fewer, and the observed *t*-value was -2.41. What would her conclusion be? The answer is that she should not reject the null hypothesis. Looking at Fig. 7.4 we see that -2.43 is in the region of retention of H_0. This is so even though in a two-tailed test she would have observed that those taking the drug had significantly fewer headaches. The point is that investigators must state their hypotheses carefully and stick to the decision rules which are consistent with the hypotheses. Stating a one-tailed or directional null hypothesis makes it easier to find differences or to reject the null hypothesis if the results are in the predicted direction. However, it also eliminates the chances of finding a significant difference in the opposite direction. Since most researchers are not willing to rule out results in either direction, two-tailed tests are used most often.

The *t*-Test For Matched Pairs

In the last example, we considered a *t*-test for two completely independent samples. The first observation for group 1 is independent from the first observation for group 2. There are many cases in which pairs of observations are related to one another, and it is of value to consider this relatedness in the analysis. Tests used to analyze these situations are called **matched pairs *t*-tests, correlated sample *t*-tests,** or **_t_-tests for dependent observations.** These terms are often used interchangeably.

Examples in which matched pairs *t*-tests are used include the following.

1. A teacher administers a reading level test the first day of the third grade and readminsters the same test at the end of the year. He wants to determine how much the children gained in reading ability over the course of the year.

2. A group of diabetic patients is on a special weight loss program designed to improve their metabolic control. Measures of metabolic control are taken before and after the weight loss program for each patient. The doctor conducting the program wants to know if the program improved metabolic control.

3. A clinical psychologist wants to know whether people are in better moods on sunny than on cloudy days. She obtained a group of individuals and had them record their moods on all days. Then for each subject she selects a sample of mood ratings from 10 cloudy days and a sample of mood ratings for 10 sunny days.

Each of these situations is one in which there are two observations for the same group of individuals.

As in the previous examples of the *t*-test, the *t*-test for the difference between paired observations is a ratio. Differences between pairs are divided by the average difference that might be expected by chance. In this case, the denominator of the *t*-formula will be the standard error of the mean for differences between observed pairs.

Box 7.3 shows the step-by-step calculation for the t-test of differences between matched samples. The example is for the difference in performance on the SAT verbal test before and after a group of students took a course preparing them for the exam.

The formula for a t-test for matched pairs is

$$t = \frac{\overline{D}}{\dfrac{S_D}{\sqrt{N_i}}} \tag{7.6}$$

In this formula, $\overline{D}$ is the mean difference between observations. The bottom half of the equation is the standard error of the difference between means.

Actually this formula is sometimes difficult to work with. Therefore, in Box 7.3 we use the mathematically equivalent formula

$$t = \frac{\Sigma D}{\sqrt{\dfrac{N \Sigma D - (\Sigma D)^2}{N - 1}}} \tag{7.7}$$

Notice from the scores in Box 7.3 that most students improved as a function of the preparatory course. However, some students actually did more poorly the second time they took the test. For instance, Jason's score was 2 points lower, and Chip's score was 14 points lower. Dustin improved by only 3 points. On the basis of these observations, it might be debated whether the course really helps. Our question is whether there is a significant average improvement in SAT scores from before the course to after the course.

To evaluate this question we must first state the null hypothesis, which is

$$H_0: \quad \mu_D \leq 0.$$

The μ_D is the population difference between pairs of observations. In this example, we are asking whether the course *improves* performance on the SAT.

• BOX 7.3

Step-by-Step Calculation of Matched Pairs *t*-Test

Question: Does a preparatory course improve performance on the SAT verbal?

Formula:
$$t = \frac{\Sigma D}{\sqrt{\dfrac{N\Sigma D^2 - (\Sigma D)^2}{N - 1}}}$$

Data: SAT-V scores for 10 students before and after the prep course.

STUDENT	SCORE BEFORE	SCORE AFTER	D (DIFFERENCE)	D^2
Megan	520	530	+10	100
Tony	611	650	+39	1521
Blake	575	602	+27	729
Jason	470	468	−02	4
Dustin	528	531	+03	9
Michelle	490	525	+35	1225
Lisa	561	577	+16	256
Allison	601	612	+11	121
Matthew	681	698	+17	289
Chip	612	598	−14	196
			$\Sigma D = 142$	$\Sigma D^2 = 4450$

Despite this example, the value of SAT preparatory courses is a matter of debate. For those interested in this controversy, see Messick, 1984.

There is little reason to think that it would decrease performance. Therefore, this is a directional or one-tailed test. So the alternative hypothesis would be

H_1: $\mu_D > 0$.

The .05 significance level is chosen. To determine the critical value of t, go to Appendix 2. For this problem, there are $N - 1$ degrees of freedom. Since there are 10 subjects in the study, there are 9 degrees of freedom. Locate 9 in the column under *df.* Now go across the table until you find the column labeled .05 for a one-tailed test. This is the second column of *t*-values from the left of the table. The value associated with 9 degrees of freedom is 1.833.

STEPS

1. Find the difference between pairs of scores for the students. Change is found by subtracting the observation on the first occasion from that on the second occasion. For Megan it would be

   $530 - 520 = +10$.

   Create a column of these differences (column labeled D).

2. Square each difference and create another column of D^2 (see right-hand column).

   $10^2 = 100$
   $39^2 = 1521$

   $\vdots$

   $-14^2 = 196$

3. Find the sum of D (or ΣD).

   $10 + 39 + \cdots + (-14) = 142$

4. Obtain the sum of D^2 (or ΣD^3).

   $100 + 1521 + \cdots + 196 = 4450$

5. Multiply ΣD^2 by N to get $N\Sigma D^2$.

   $10(4450) = 44500$

6. Calculate $(\Sigma D)^2$ by squaring the results of Step 3.

   $142^2 = 20146$

   Caution: Be aware of the difference between D^2 and $(D)^2$.

7. Subtract Step 6 from Step 5 to get $N\Sigma D^2 - (\Sigma D)^2$.

   $44500 - 20146 = 24354$

8. Divide Step 7 by $N - 1$.

   $\dfrac{24354}{9} = 2706$

9. Take the square root of Step 8.

   $\sqrt{2706} = 52.02$

   This is the denominator of the *t*-formula.

10. The *t*-value is the ratio of Step 3 to Step 9.

    $t = \dfrac{142}{52.02} = 2.73$

11. The degrees of freedom are

    $N - 1 = 10 - 1 = 9$.

12. Evaluate the null hypothesis (see text).

Using this information, the decision rule can be stated as follows:

If the observed t is less than 1.833, retain H_0.

If the observed t is greater than or equal to 1.833, reject H_0 and accept H_1.

The example in Box 7.3 shows that the observed t-value is 2.73. According to the decision rule the null hypothesis should be rejected and the alternative accepted. This suggests that in repeated samples of size 10, the probability of observing improvement as great as that found in the sample by chance is less than 5 in 100. Thus, the improvement is not considered to have happened by chance.

When interpreting a paired t-test, it is often important to consider all alternative explanations for the change. In the example of SAT scores, other factors (not the prep course) may be responsible for the improvement in test performance. For instance, it is possible that only the most motivated students would enroll for the preparatory course. The improvement might reflect the motivation of the students rather than the benefits of the course. Typically it is difficult to say that a treatment caused a change unless similar comparisons are made for a control group. You can learn more about these problems in courses on experimental psychology and experimental design.

Types of Error

When the null hypothesis is rejected, we conclude that the observed differences were improbable by chance alone. In the previous examples we concluded that the difference in scoring between the San Diego Chargers and the New England Patriots was unlikely to have occurred as a result of chance or sampling error. In the other example we found that the improvement in SAT scores after the preparatory course was most likely not explainable simply by sampling error. The key word in each of these statements is *probably*. When making these statements, we say that the odds are on our side. However, what if we are wrong?

In statistical inference there is no way to say with certainty we have made the right choice by retaining or rejecting the null hypothesis. There are two types of errors we might make. First, we might reject the null hypothesis when indeed we should not reject it. This is called a **type I error.** Second, we might retain the null hypothesis when indeed it should be rejected. This is called a **type II error.** Let's consider these decision errors in more detail.

■ Type I Error

A type I error occurs when we reject the null hypothesis when in fact, the null hypothesis should be retained. This might occur when a researcher decides that two means are different. She might conclude that the treatment works or that groups are not sampled from the same population. Yet in reality the observed differences are due only to sampling error. Science is very conservative, and we do not want to make type I errors very often. This would lead us to advocate treatments that actually do not work. Consider the case of a drug experiment in which the researcher concluded the new medication works significantly better than a placebo. Yet this conclusion was a type I error. This is a mistake most reasearchers would not want to make.

Because of researchers' desire to avoid type I errors, statistical models have been created so that the investigator has control over the probability of a type I error. At the .05 significance or alpha level, a type I error occurs in 5% of all cases. At the .01 level it occurs in 1% of all cases. Thus, at the .05 alpha level, one type I error will be made in each 20 experiments. At the .01 alpha level one type I error will be made in each 100 experiments. Consider the situation of a class of students each conducting their first experimental project. Imagine that each of these projects is independent of one another and that none of the experimental interventions for the entire class actually works. Among 20 students in the class, we would expect one to conclude that a treatment worked even if the results were actually determined randomly. If you conduct the same experiment 20 consecutive times, you would expect to find a significant result on one occasion simply by chance.

■ Type II Error

Making a type I error may be embarrassing and something we want to avoid. Yet attempting to avoid a type I error might increase the chance of making a type II error. In this case, we retain the null hypothesis when in fact it is actually wrong. In other words, the treatment actually works, but we conclude it does not. The probability of a type II error is symbolized by the Greek capital letter beta, β.

Of course, we do not always make errors. Table 7.1 shows four possible outcomes broken down in a two-by-two table. In the left column we have the decisions the researcher made about the null hypothesis. She can either reject H_0 or not reject H_0. Across the top of the table we have the real world or actual situation. The researcher does not know this, but hypothetically we can say this is the decision that "should have" been made. Sometimes we refer to this as "God's model," implying that there is a correct decision that could have been made but it is beyond our capability of knowing.

Now consider the entries in the table. First there is the upper left box that shows that the null hypothesis is rejected even though it is true. This is a type I error. It will occur with a probability of alpha (α), or the significance level for the test. For example, at the .05 alpha level, the probability of this type of error is 5 in 100. Next look at the other type of error which is the bottom right box in the table. Here the decision was not to reject the null hypothesis when in the real world the null hypothesis was false. This is a type II error with the probability of beta (β).

Table 7.1			
Four possible outcomes of decisions concerning the null hypothesis			
		"GOD'S MODEL" WHAT ACTUALLY IS TRUE	
		H_0 is True *It should have been retained*	H_0 is False *It should have been rejected*
RESEARCHER'S DECISION	Reject H_0	Type I error $p = \alpha$	Correct decision $p = 1 - \beta$ (also called the power of the test)
	Retain H_0	Correct decision $p = 1 - \alpha$	Type II error $p = \beta$

Of course, not all decisions are incorrect. The table also shows two boxes where the researcher made the correct decision. First there is the lower left box in which the decision was to retain the null hypothesis when indeed the null hypothesis was true. This is a correct decision, and it occurs with a probability of $1 - \alpha$. Finally there is the situation portrayed in the upper right box. Here the investigator decided to reject the null hypothesis when indeed the null hypothesis was false. This is a correct decision with the probability $1 - \beta$. The upper right cell is of particular interest. It is often called the *power of the test*. The power of a test is discussed in the next section. Before turning to this topic, it is important to relate the concepts of type I and type II errors to the probability sampling distribution.

Figure 7.5 shows sampling distributions of sample means under the null hypothesis and the alternate hypothesis. The null hypothesis is the distribution on the left labeled H_0; the alternative hypothesis is the one on the right labeled H_1. Under the null hypothesis, the probability of a type I error is equal to α. In this case it is a two-tailed test, so the area for alpha is divided between the upper and lower tails of the distribution. The rest of the distribution under H_0 is the probability of a correct decision, which is $1 - \alpha$. Now suppose that the alternative hypothesis H_1 is correct. The distribution of sample means under H_1 is shown in the right distribution in the figure. Imagine that even though H_1 is correct, we are going under the assumption that H_0 is correct. How much of the distribution of H_1 falls within the region of retention of H_0? This is a complicated question, so take some time to think it over. Looking at the figure, you can see that the two distributions overlap. Look at the critical value for the upper region of the confidence interval for H_0. Any score below that (but above the lower critical boundary) will be within the region of retention for H_0. In other words, even though the alternative hypothesis is correct, if a sample

---

Figure 7.5

Correct decisions, type I, and type II errors in relation to sampling distributions.

The outside color areas represent the probability of a type I error, and the inside color areas represent the probability of a type II error.

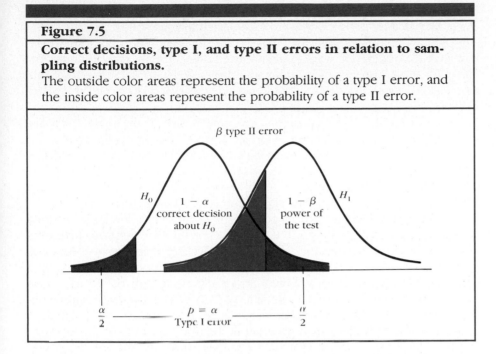

---

mean falls below the upper critical boundary under the null hypothesis, the null hypothesis will be retained. These situations define type II errors. Now examine the area under H_1 above the upper critical boundary for the null hypothesis. This area represents the power of the test. The power of the test is a very important concept, which we will now discuss.

Power of the Test

The conservative nature of statistical inference requires that we be cautious about making errors. If we are very conservative and set the probability of a type I error very low, there may be serious consequences. In particular, by avoiding type I errors we may increase the likelihood of making a type II error. Thus, by manipulating the alpha level we can reduce the probability of type I error but may increase the probability of type II error.

Actually, there are several things an experimenter can do to gain control over the probability of different types of errors and correct decisions. One type of correct decision that is particularly important is the probability of rejecting the null hypothesis and being correct in that decision. *Power* is defined as the probability of rejecting the null hypothesis when, in the real world, it should be rejected. This is different than the probability of a type I error which is the probability of rejecting the null hypothesis by chance. Researchers like to have the power of their statistical tests high because it increases their chances of being right about the decisions they make concerning the null hypothesis. There are a variety of different ways to manipulate power. The theoretical underpinnings for these methods of manipulating power are beyond the scope of this book. However, in practice we should consider several relationships that affect power.

One factor that affects power is the sample size. If the sample size increases, power increases. In other words, the larger the sample the greater the probability that you will make the correct decision when the results lead you to reject the null hypothesis. Statisticians have worked out tables that show the relation between sample size and power. Many researchers use these tables to decide how many subjects they will need to conduct an experiment.

Another factor that influences power is the significance level. As the alpha or significance level increases, the power increases. For instance, if the .05 level is selected rather than the .01 level, you have a greater chance of rejecting the null hypothesis. However, you also have a higher probability of making a type I error. This situation is shown graphically in Fig. 7.6. The figure shows the relation between power and alpha level. The left portion of the figure considers the situation when the alpha level is .20. Here 10% of the distribution under the null hypothesis is in each tail of the distribution. Note that the area for the probability of a type II error is relatively small, and the area for power is relatively large. The center section of the figure shows the situation for an alpha level of .10, that is, that the probability of a type II error increases while the error for the power of the test decreases. The right panel of the figure shows the situation for an alpha level of .01. Here notice that the area for a type II error is much larger than any other section, and the area for the power is smaller. The figure demonstrates that there is a trade-off between a selection of an alpha level and power. By reducing our chances of making a type I error, we also reduce the chances of correctly identifying a real difference (power).

Figure 7.6

Relationship between power and α level.

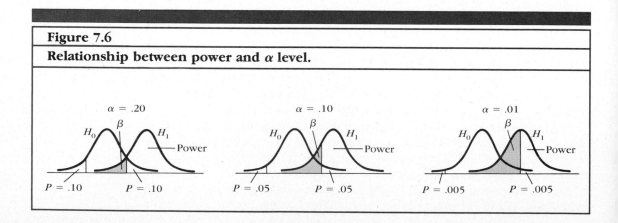

Thus, the safest manipulation to affect power without affecting the probability of a type I error is to increase the sample size.

It is worth noting that not all statistical tests have equal power. Some statistical methods are "more powerful" than others. This means the probability of correctly rejecting the null hypothesis is higher with some statistical methods than with others. For example, nonparametric statistics, which will be discussed in Chapter 12, tend to be less powerful than *parametric* tests discussed in this and the next two chapters. Chapter 12 reviews some of the important differences between parametric and nonparametric statistics.

Summary

The *t*-test is a common statistical procedure used to compare two means. There are several types of *t*-tests. One type compares an observed mean from a sample of observations with a hypothetical population mean. Another type compares the means of two independent samples or independent groups of subjects. A third type compares means from two different sets of observations of the same individuals or cases.

Statistical tests, such as the *t*-test, are subject to several types of errors. A type I error occurs when an experimenter incorrectly rejects the null hypotheses and concludes that two means are different. The error occurs when the null hypothesis should have been retained.

A type II error occurs when an experimenter or researcher retains the null hypothesis (that the two means do not differ) when it should have been rejected. In other words, the means really differed, but the researcher concluded that they were not significantly different. Subtracting the probabability of a type II error from 1.0 gives the power of the test. This is the probability of finding a significant difference, if indeed there was one to be found. The power of the test can be improved by increasing the sample size and reducing measurement error and variance. The power also tends to be larger as true differences between means increase.

Key Terms

1. **Two-tailed Test** A nondirectional test of the null hypothesis. In a two-tailed or nondirectional test, a significant difference can be either above or below the hypothetical population mean. For example, with a two-tailed test at the .05 significance level, the null hypothesis would be rejected if the observed mean for difference between means was in the top 2.5% or the bottom 2.5% of the sampling distribution of sample means.

2. **One-tailed Test**   A directional test of the null hypothesis. With a one-tailed test, the experimenter states the specific end of the sampling distribution that should be used for the region of rejection of the null hypothesis. For example, an experimenter studying weight loss may state that Group B should lose more weight than Group A. Thus, the null hypothesis would be rejected only if it was statistically improbable that the amount of weight Group B lost was greater than Group A. If Group A lost more weight than Group B, the null hypothesis would not be rejected.

3. **Standard Error of the Difference**   The standard error in estimating the difference between sample means.

4. **Paired *t*-Test**   A *t*-test used to compare the difference between two sets of observations made for a common group of cases. For example, if a group of students was measured at two points in time, a paired *t*-test may be used to compare the difference between these two sets of observations on the same individual.

5. **Type I Error**   The probability of rejecting the null hypothesis when in fact it should have been retained. The probability of a type I error is under the control of the experimenter and is usually called the alpha or significance level.

6. **Type II Error**   An error made when the null hypothesis is retained when in fact it should have been rejected. This is sometimes called a false negative. The probability of a type II error is symbolized as beta (β) and is only indirectly under the control of the experimenter. The experimenter can reduce the probability of a type II error by increasing the sample size.

7. **Power of the Test**   The probability of rejecting the null hypothesis when indeed the null hypothesis should be rejected. The power of the test is calculated by subtracting β (or the probability of a type II error) from 1 $(1 - \beta)$. Thus, the power of the test is the probability of correctly identifying the significant differences.

Exercises

1. The average IQ score on the Stanford-Binet Intelligence Test is 100 with a standard deviation of 15. Ten children were randomly selected from the Palo Alto School District. Their IQ's were as follows: 106, 116, 97, 141, 122, 127, 119, 113, 122, and 106. Was this a representative sample from the general population?

2. A group of dogs was randomly assigned to eat dog food A or dog food B. Both dog foods were chunky style, and each chunk was of equal size. Each

dog was given 100 chunks of food, and the experimenter counted the number of chunks left in the bowl for each dog. The number of chunks left were as follows: Group A: 7, 11, 4, 14, 21, 0, 6, 4, 10, 9; Group B: 0, 1, 3, 0, 2, 7, 2, 8, 6, 1. State the null hypothesis and the alternative hypothesis of the experimenter. Should the null hypothesis be retained or rejected? How should the experimenter state his conclusion?

3. A group of 16 women has joined a weight-loss clinic. The clinic manager randomly chooses half of the women for a special program in which they receive a dollar of their initiation fee back for each pound they lose. He is interested in whether this incentive produces *significantly more* weight loss. State his null hypothesis, alternative hypothesis, and decision rule.

4. Ten high school runners practice the 100-yard dash with and without their sweatpants. The coach wants to determine whether running in sweatpants slows performance. (He counterbalances the experiment so half of the runners run first with their sweatpants and second without their sweatpants, and the other half do the reverse.) The times are as follows.

RUNNER	WITH SWEATPANTS	WITHOUT SWEATPANTS
John	10.7	10.3
George	9.0	9.7
Fred	12.6	11.4
Hal	11.4	10.7
Josh	11.6	11.3
Howard	13.1	13.6
Bill	12.4	11.9
Ted	10.7	10.6
Jerry	12.4	12.0
Cathie	10.2	9.7

State the null hypothesis, the alternative hypothesis, and the decision rule. Did the athletes run significantly faster without their sweatpants?

5. In an experiment with a .05 significance level, what is the probability of a type I error?

6. What is the relationship between a type I error, a type II error, and the power of a test?

7. If you were evaluating a new, expensive medical procedure, would it be important to have a high power for the test? Why?

8

The One-Way Analysis of Variance

The last chapter covered techniques for comparing two means. More precisely, it covered methods for examining two sample means and asking the probability that they were drawn from the same population. These techniques are applicable to a wide variety of experiments that compare two treatment groups or one treatment group with a control group.

There are many research problems that require that we compare more than two groups. Analysis of variance (or ANOVA) is a statistical method for comparing two or more groups of observations. Consider the following examples.

1. A clinical psychologist wants to evaluate the benefits of behavior therapy for weight loss. She decides to use an experimental group, which receives behavior therapy, and two different control groups. One control group receives an equal amount of attention as the group participating in the behavior therapy. The second control group receives neither attention nor behavior therapy. Weight loss is the dependent variable, and the three groups are compared using a one-way analysis of variance.

2. A health scientist wants to evaluate the effects of 6 different approaches to health education. While working in a school district with 60 schools, he randomly draws samples of 10 schools to receive each intervention. Change in health behavior is the dependent variable, and a one-way analysis of variance is used to test for differences across the 6 treatments.

In each of these examples we could perform multiple t-tests. In the first example there were three groups. To make the comparisons three t-tests would be necessary (experimental vs. control 1, experimental vs. control 2, and control 1 vs. control 2). However, as the number of groups increases the number of tests also increases. So in the example of the health scientist comparing 6 treatments the number of t-tests required to make all of the comparisons would be 15.*

When many tests are made we run into the **multiple comparisons problem:** When multiple comparisons are made the probability of finding at least one difference by chance increases. At the .05 significance level we would expect to find one significant difference by chance in each 20 trials (that is, in 5% of all comparisons). The number of possible comparisons is equal to

$$\frac{J(J-1)}{2},$$ (8.1)

where J is the number of groups. In the example of 6 groups the number of comparisons is

$$\frac{6(6-1)}{2} = \frac{30}{2} = 15.$$

As you can see, with this many comparisons it is likely that at least one significant difference will be observed by chance alone. Thus, t-tests are not appropriate when making statistical comparisons among more than two groups.

The method introduced in this chapter is not designed to test individual differences between treatment means. Instead, it asks whether there are differences across the different treatments without necessarily considering differences between individual pairs. Often experimenters and scientists want to know if, in general, there are differences among groups. Then they use other procedures to identify specifically which groups differ from one another. However, the first step is to determine whether there are any differences among the treatment groups. Analysis of variance is the method most often used to test for differences between three or more group means. It can also be used to test for differences between two group means, and in this case we get exactly the same information as the t-test would give us. In this chapter we examine analysis of variance procedures that consider one independent variable at a time. In the next chapter we present methods that examine two different independent variables or factors in the same analysis.

The analysis of variance procedures were developed by the eminent British statistician Sir Ronald A. Fisher. Box 8.1 gives a brief biographical sketch of Sir Fisher.

---

*For groups 1, 2, 3, 4, 5, and 6, the possible comparisons are 1 vs 2, 1 vs 3, 1 vs 4, 1 vs 5, 1 vs 6, 2 vs 3, 2 vs 4, 2 vs 5, 2 vs 6, 3 vs 4, 3 vs 5, 3 vs 6, 4 vs 5, 4 vs 6, and 5 vs 6.

• BOX 8.1

Sir Ronald A. Fisher

Ronald A. Fisher is perhaps the most important figure in the history of modern statistics. As a young man Fisher studied traditional mathematics. When he applied for graduate school at Cambridge, his letters of recommendation noted that he was an intelligent young man, but his methods were somewhat unorthodox and did not conform to the standards of British mathematicians. Nevertheless, Fisher was admitted to the program, and he later began developing his statistical models.

Much of Fisher's early work was in the field of agronomy. In the 1920s and 1930s, many experiments were being devised to find the most profitable methods for farming. Experiments were set up in which different treatments were applied to different fields in order to find the best way to obtain large yields of crops. In advanced statistics courses you may come across special methods of analysis of variance called split-plot designs. This name comes from Fisher's description of experimental designs in which plots of land were split in half and different treatments were applied to the different sections of land.

Besides the analysis of variance, Fisher made many other contributions to statistical science. He invented a variety of statistical techniques and expanded older methods for a wide range of new problems. In addition, he also had a suspicion of other scientists. For example, Fisher used the laws of probability to imply that other scientists had faked their data. In one case, he demonstrated that the famous geneticist, Gregor Mendel, had probably not been completely honest in reporting the results of his experiments with peas. Although Mendelian genetics has continued and Mendel was probably right in his inferences, the chances that he could have obtained the results he reported in a real experiment are extremely small.

Sir Ronald A. Fisher
Eminent statistician, scientific muckraker, proud father.

In addition to being an eminent scientist and statistician, Fisher was the father of nine children. According to his son-in-law, also an eminent statistician named G. E. P. Box, Fisher kept records of the growth of his children in family bibles. The growth patterns were analyzed in the bibles using a variety of statistical methods.

The Rationale for the Analysis of Variance

Experiments involving comparisons across many groups really ask the question, "Are the sample means for the different groups drawn from the same or from different populations?" The question is similar to that asked in the t-test, where two sample means are drawn and compared. A significant difference between sample means is the basis for inferring that the samples were drawn from different populations. Now we consider more than two means. This time, instead of evaluating the difference between means, we consider the *variance* across the means. For example, suppose that we have done an experiment on weight loss that compares behavior therapy to an attention control group and a no-treatment control group. The mean number of pounds lost in the behavior therapy group was 12 and the pounds lost in the attention and no-treatment control groups were 2.6 and .8, respectively. In the analysis of variance we actually estimate the variance across these three means. The more the three treatments differ from one another, the greater will be the variance across the different sample means. If the groups do not differ, then variance among the sample means will be small.

How do we determine whether observed variability between sample means is small or large? We have to determine the extent to which we would expect the sample means to vary by chance. One way of doing this is to find the variability within each of the treatment groups. This would be an index of how much variability we would expect on the particular outcome measure. Since individuals within each of the treatment groups are treated exactly the same, variability within the groups could not be attributed to the differences between treatments.

Table 8.1 summarizes data from a hypothetical weight-loss experiment. The table describes three groups: behavior therapy, attention control, and no-treatment control. Within each group, a subject-by-subject breakdown of weight loss is given. Notice that in the behavior therapy group all of the subjects lost weight. That is why there is a negative sign for each of the numbers in pounds lost. In the attention control group, most of the subjects lost weight; however, one subject (Mike) neither lost nor gained weight while another subject (Tim) gained weight. Check the table for Mike and Tim and you will notice that neither of them has a negative sign. In group 3, the no-treatment control group, several of the subjects gained weight (Tod, Paul, Dan, and Dawn). One subject (Jill) neither lost nor gained weight. At the bottom of the table are listed the means, standard deviations, and variances for the different groups. The means for the three groups are -12 for the behavior therapy group, -2.6 for the attention control group, and $-.8$ for the no-treatment control group.

This example, which uses both positive and negative numbers, will be

Table 8.1

Example data from weight-loss experiment: pounds lost over 10 weeks*

$$H_0: \quad \mu_1 = \mu_2 = \mu_3$$
$$H_1: \quad H_0 \text{ is not correct}$$

GROUP 1 BEHAVIOR THERAPY		GROUP 2 ATTENTION CONTROL		GROUP 3 NO TREATMENT	
Subject	Pounds Lost	Subject	Pounds Lost	Subject	Pounds Lost
X_{11} Joe	−6	X_{12} Mike	0	X_{13} Tod	2
X_{21} John	−8	X_{22} Tom	−2	X_{23} Tony	−4
X_{31} Fred	−21	X_{32} Harry	−1	X_{33} Terry	−6
X_{41} Irwin	−17	X_{42} Tim	3	X_{43} Paul	3
X_{51} Blake	−16	X_{52} Howard	−6	X_{53} Dan	1
X_{61} Sally	−14	X_{62} Nancy	−1	X_{63} Dawn	4
X_{71} Carie	−11	X_{72} Cathie	−4	X_{73} Jan	−5
X_{81} Torey	−9	X_{82} Phillis	−1	X_{83} Debbie	−2
X_{91} Allison	−8	X_{92} Connie	−2	X_{93} Jill	0
X_{101} Blanche	−10	X_{102} Michelle	−12	X_{103} Megan	−1
$\Sigma X_{.1} = -120$		$\Sigma X_{.2} = -26$		$X_{.3} = -8$	
$\bar{X}_{.1} = -12$ $S_{.1} = 4.81$ $S_{.1}^2 = 23.11$		$\bar{X}_{.2} = -2.6$ $S_{.2} = 4.06$ $S_{.2}^2 = 16.49$		$\bar{X}_{.3} = -.8$ $S_{.3} = 3.43$ $S_{.3}^2 = 11.73$	

used throughout this chapter. Analysis of variance can be used for positive and negative numbers. However, the results will always be in positive numbers. The example should provide a good challenge for working with negative numbers.

It is possible for us to calculate a variance across the three observed means. Recall from Chapter 3 that the variance can be obtained from the formula

$$S^2 = \frac{\Sigma(X - \bar{X})^2}{N - 1},$$

or

$$S^2 = \frac{\Sigma X^2 - \dfrac{(\Sigma X)^2}{N}}{N - 1}. \tag{3.5}$$

If you calculate the variance using this formula (you don't need to do this now) you will get a variance of 36.17. The question inherent to the analysis of variance is whether this variability across the group means is significant.

In summary, the logic of the analysis of variance is to find the two independent estimates of variability. One is the variance across group means. This is actually the variance of the differences among groups. The other is the variance within groups, which determines how much scores vary by chance on the dependent variable of interest. The analysis of variance test is a ratio of these two variances.

Variance Within Groups

In the following pages we offer a more technical explanation of the analysis of variance. This explanation requires that you understand the notion of **partitioning.** In partitioning, we take the total variation in an entire experiment and divide it into components. The total variation, or total variance, would be the variance for the entire aggregate of scores regardless of group assignment. In addition to the total variance, there is also a total or *grand mean*. As we will show, the total variance can be divided into two components: **between-groups variance,** which is the variance of group means around the grand mean, and **within-groups variance,** which is the variance of individual scores around their own group mean.

Perhaps this will be easier to understand if we look back to Table 8.1. For the entire experiment described there, the *grand mean,* or the mean across all treatments, is −5.13. This means that the average subject in the experiment lost 5.13 pounds. Now let's take an individual, such as Blanche, who lost 10 pounds (see the 10th person in group 1.) Notice that Blanche's weight loss is not the mean or average weight loss. Blanche lost 10 pounds while the mean loss was 5.13 pounds. Thus, there is difference of 4.87 pounds between Blanche and the grand mean. This difference can be broken down into two components. The first component is the difference between Blanche's group mean and the grand mean. Blanche is a member of the behavior therapy group which lost an average of 12 pounds. Thus, her group differs from the grand mean by 6.87 pounds:

$$-12 - (-5.13) = -6.87.$$

The second component of the variability of Blanche from the grand mean is her variability from her own treatment mean. Blanche lost 10 pounds and members of her group, on the average, lost 12 pounds, so the difference between Blanche and her group mean is 2 pounds: $-10 - (-12) = -2$. Thus her difference from the grand mean can be stated as

$$6.87 - 2 = 4.87.$$

In the next section, this notion of partitioning will be extended to the formal model of the analysis of variance.

Separate Variance Estimates

To understand the logic of the analysis of variance, think back to our discussion of sampling theory in Chapter 5. Suppose that we draw random samples from a large population. Remember that each number that we draw from the large population is an unbiased estimate of the population mean. If we draw a sample of ten observations and calculate the mean of the sample, this will also be an estimate of the population mean. The variance of the sample will be an estimate of the population variance.

Now suppose we take four random samples, each with ten observations. Next we calculate the mean for each of the samples. What would we expect the variance of the means to be? Since each sample mean is an unbiased sample of the population mean, the variance across the four means should be an estimate of the population variance. Similarly, the variance within each of the samples should also be an estimate of the population variance.

It can be proven mathematically that if observations are drawn randomly from the same population, the expected variance across the four means will be the same as the expected variance within each sample. It is possible that the sample means will not be the same as one another. The question we are interested in is whether we should attribute the difference we observe in sample means to chance or to some systematic experimental effect.

In an experiment, we assume that each member of each treatment group has been treated in the same way. Thus, differences between cases within the same group are attributed to sampling error. The within-group variation, pooled across all groups in the analysis of variance, provides the estimate of how much we would expect scores on the dependent variable to differ from one another.

In summary, the analysis of variance uses two separate estimates of variability. One estimate is obtained from variability across cell means. The other estimate is based on variability within treatment groups. Only the between-groups variability will be affected by treatment. Remember that the within-groups variability only considers differences between individual cases and their cell mean. Within any cell, all cases receive the same treatment. An estimate of the effect of treatment in analysis of variance is provided by the ratio

$$\frac{\text{Between-groups variability}}{\text{Within-groups variability}}$$

If differences between cell means represent only sampling error, the top and bottom portions of this ratio are independent estimates of sampling error. Thus, the expected value for this ratio would be 1.0. However, if the differences between cell means are greater than would be expected on the basis of sampling error, then the ratio should be larger than 1.0. Remember that the two portions of this ratio are independent of one another. Each is based on a different number of observations. The top portion, or between-groups portion, of the ratio is based on the number of groups. The bottom is based on the number of subjects within each cell and summed across the cells. Degrees of freedom for the top half are the number of groups minus 1. For the bottom, the degrees of freedom are obtained from the number of subjects in each cell minus 1 and summed across the number of cells. For example, if there were ten subjects in each of three cells, the degrees of freedom for the bottom portion would be

$$(10 - 1) + (10 - 1) + (10 - 1) = 27.$$

The null hypothesis in the one-way analysis of variance is that all population means are equal for j-groups, or

$$H_0: \mu_1 = \mu_2 = \mu_3 \ldots = \mu_j.$$

This is tested against the alternative hypothesis, or

$$H_1: H_0 \text{ is not true.}$$

The evaluation of the null hypothesis in the analysis of variance, then, is based on a ratio of two independent estimates of variance. Each of these estimates is associated with its own degrees of freedom. To evaluate the difference, we use a distribution known as the F-distribution, which is shown in Appendix 3. Take a minute to study Appendix 3. The F-distribution is different from the t-distribution in several ways. First, there are no negative values in the F-distribution. This is because the analysis of variance is a ratio of variances. Recall from Chapter 3 that it is impossible to have a negative variance since the variance is based on *squared* deviations from the mean. Thus, it would be theoretically impossible to have a ratio of variances that would produce a negative sign. Second, the F-values are associated with two different degrees of freedom: one for the numerator and one for the denominator. To find a critical value for F you must look in the table using information about both the numerator and the denominator. Take, for example, a problem with 2 degrees of freedom for the numerator and 27 degrees of freedom for the denominator. First, go down the column labeled "n_2" until you reach 27. This column is for the degrees of freedom associated with the denominator. Then move across the row to the right to the second column, which is associated with 2 degrees of freedom for the numerator. There are two tabled values there: one for the .05 significance level and one for the .01 significance level. The .01 values are listed in the second row of each pair. Let's suppose we had chosen the .05 significance level. The critical value then is 3.35. Now try another example with 5 and 34 degrees of freedom. Go down the first column until you reach 34. Then go across the row until you hit the fifth column of numbers. These are the critical values for the F-statistic with 5 and 34 degrees of freedom. The table shows they are 2.49 for the .05 significance level and 3.61 for the .01 significance level.

Assumptions

The use of the analysis of variance depends on several assumptions. The most important assumptions are the following:

1. The populations from which samples are drawn are normally distributed.
2. The cases for the study have been drawn randomly and independently from the populations.
3. The variances of the populations from which cases are drawn are equal.

The analysis of variance is known to be relatively robust to violations of these assumptions. By *robust* we mean relatively unaffected. However, severe violations of these assumptions would mean that the probability values associated with the F-statistics in Appendix 3 are not accurate. There are tests that can be used to determine whether assumptions such as "equal population variances" are met. However, presentation of those tests is beyond the scope of this book. When a researcher is concerned that he or she may have violated these assumptions, alternative methods of data analysis are available. One of these methods, the Kruskal–Wallis test, is presented in Chapter 12.

Analysis of variance is in itself a topic of several advanced statistics courses. In this book only the most basic methods are presented. It is more difficult to be confident that we have met the assumptions if the sample sizes for different groups are not equal. Therefore, we focus on problems with equal numbers of subjects in each group. Solutions for the analysis of variance with unequal group sizes are usually presented in more advanced courses.

The Computation of the One-Way Analysis of Variance

One of the obstacles for many students of statistics is that there is no common notation system. Different books use different notations. As a result, students often learn formulas in one course only to find that in the next course the notational system is different. For this section, we use the notational system common in advanced analysis of variance texts. That notational system is summarized in Box 8.2. In the one-way analysis of variance, each case is identified by two subscripts. The first of the subscripts is for the subject number and the second is for the group number. For example, X_{43} would identify the score for subject 4 within group 3. It is important that you study Box 8.2 before attempting to interpret the analysis of variance formulas.

• BOX 8.2

Introduction to Dot Notation

Each data point is indexed by two subscripts. The first is the subject number. For subjects 1 through 10, these are

1, 2, 3, 4, 5, 6, 7, 8, 9, and 10.

The second subscript is the group number. If the experiment has three groups, these would be 1, 2, and 3.

An example is X_{32}. This refers to the score for the third subject in group 2. Referring back to Table 8.1, X_{32} is the score for Harry.

An index that has been summed is replaced with a dot. For example,

$\Sigma X_{.2}$

means we have summed over the individuals in group 2. Referring to the data in Table 8.1, it means

$0 + (-2) + (-1) + 3 + \cdots + (-12)$.

The notation

$\Sigma X_{..}$

suggests that both indexes have been summed over.

The notation

$\overline{X}_{.1}$ is the mean of group 1,

$\overline{X}_{.2}$ is the mean of group 2, and

$\overline{X}_{.3}$ is the mean of group 3.

• Calculation of the Analysis of Variance

The components of the analysis of variance are described in Table 8.2. There are several important components to be considered. There is the total sums of squares, which is obtained by summing the squared deviation of each individual score around the grand mean. Remember that the grand mean for the example given in Table 8.1 was -5.3. The total sum of squares could be obtained by subtracting this value from each case in Table 8.1, squaring the difference, and summing the squared differences across all individuals. In practice, one rarely calculates the total sum of squares directly. Rather, the total sum of squares is derived from summing its two components, the between-groups sum of squares and the within-groups sum of squares.

The between-groups sum of squares represents the deviation of the cell means around the grand mean. For example, the cell mean for group 1 was −12 and the grand mean was −5.13. Thus, the deviation around the grand mean for this cell is −6.87. This value squared is 47.19. Since there were ten subjects in this cell, we multiply this squared deviation by 10. This calculation is performed for each of the cells, and the result is summed across cells to obtain the between-groups sum of squares. Notice that the number of cases in each group or cell is denoted with a lower-case n while the total number of cases in the analysis is signified with an upper-case N.

The within-groups sum of squares is obtained by finding the squared deviations of individual scores around their own group mean. For example, the mean of group 1 was −12. The first subject had a score of −6. So we take

$$-6 - (-12) = 6,$$

and square it to obtain 36. Then we take the next subject, who had a score of −8, subtract −12 to obtain 4, and square it to get 16. This is done for all of the other subjects in group 1. Then we go on to group 2 where the cell mean was −2.6. The first subject had a score of 0, so we take

$$0 - (-2.6) = 2.6,$$

square it to obtain 6.76. The same calculation (subtracting −2.6 and squaring the results) is performed for all the other subjects in group 2. Finally we move on to group 3. The mean for this group was −.8. The first subject in group 3 gained 2 pounds, so the first squared deviation is obtained by

$$2 - (-.8) = 1.2.$$

Squaring this value gives 1.44. The next subject in group 3 lost 4 pounds, so the squared deviation is

$$-4 - (-.8) = -3.2.$$

This value squared is 10.24. When each of the individual scores minus its own cell mean has been obtained and squared, the values are then summed across all groups to obtain the within-groups sums of squares.

Next we must find the degrees of freedom for both the between-groups component and the within-groups component. The between-groups degrees of freedom are obtained by subtracting 1 from the number of groups, or

$$df_B = J - 1 = \text{number of groups} - 1,$$

where J is the number of groups. In this example, there were three groups, so

$$df_B = 3 - 1 = 2.$$

The within-groups degrees of freedom are obtained by subtracting the number of groups from the total number of subjects, or

$$df_W = N - J = \text{number of cases minus number of groups.}$$

Table 8.2			
Components of the analysis of variance			
STATISTICAL TERM	VERBAL DESCRIPTION	COMMENT	EXAMPLE FROM TABLE 8.1*
$SS_T = \Sigma(X_{ij} - \bar{X}_{..})^2$ $= \dfrac{\Sigma X_{ij}^2 - (\Sigma X_{ij})^2}{N}$	Total sum of squares	The sum of the individual squared deviations around the grand mean	$(6 - 5.3)^2 + (8 - 5.3)^2 + \cdots +$ $(0 - 5.3)^2 + (1 - 5.3)^2 = 1185.47$
$SS_B = \Sigma N_j(\bar{X}_{.j} - \bar{X}_{..})^2$ $= \dfrac{(\Sigma X_{.1})^2}{n_1}$ $+ \dfrac{(\Sigma X_{.2})^2}{n_2} +$ $\cdots + \dfrac{(\Sigma X_{.j})^2}{n_j}$	Between-groups sum of squares	The sum of square deviations of group means around the grand mean.	$10(12 - 5.13)^2 + 10(2.6 - 5.13) +$ $10(.8 - 5.13) = 723.47$
$SS_w = \Sigma(X_{i1} - \bar{X}_{.1})^2 +$ $\Sigma(X_{i2} - \bar{X}_{.2})^2 +$ $\Sigma(X_{ij} - \bar{X}_{.j})^2$	Within-groups sum of squares	The sum of the squared deviations of individual scores around their own group mean.	$(6 - 12)^2 + (8 - 12)^2 + \cdots +$ $(10 - 12) + (0 - 2.6)^2 +$ $(2 - 2.6)^2 + \cdots + (12 - 2.8)^2 +$ $(-2 - .8)^2 + (4 - .8)^2 + \cdots +$ $(1 - .8)^2 = 462$
$df_B = J - 1$	Between-groups degrees of freedom	Number of groups minus 1	$3 - 1 = 2$
$df_w = N - J$	Within-groups degrees of freedom	Number of subjects (total) minus number of groups	$30 - 3 = 27$

In this example, there were 30 subjects and 3 groups, so the degrees of freedom are

$$df_w = 30 - 3 = 27.$$

Finally, it is appropriate to calculate the total degrees of freedom. The total degrees of freedom are obtained by subtracting 1 from the total number of cases, or

$$df_T = N - 1 = \text{cases} - 1.$$

The total degrees of freedom should equal the sum of the degrees of freedom for between-groups and the degrees of freedom for within-groups. If these do not add up, recheck your work.

The next step in the calculation of the analysis of variance is to obtain the mean square values. The mean square values are always calculated by dividing

Table 8.2 (*continued*)

Components of the analysis of variance

STATISTICAL TERM	VERBAL DESCRIPTION	COMMENT	EXAMPLE FROM TABLE 8.1
$df_T = N - 1$	Total degrees of freedom	Total number of subjects minus 1. Also the same as $df_B + df_w$.	$30 - 1 = 29$
$MS_B = \dfrac{SS_B}{df_B}$	Mean square between groups	The ratio of between-groups sums of squares over between-groups degrees of freedom	$723.47/2 = 361.73$
$MS_w = \dfrac{SS_w}{df_w}$	Mean square within groups	The ratio of within-groups sums of squares over within-groups degrees of freedom	$462/27 = 17.11$
$F = \dfrac{MS_B}{MS_w}$	F ratio	The ratio of mean square between-groups divided by mean square within groups	$361.73/17.11 = 21.14$

*All values have been multiplied by -1 to make calculations easier.

the sum of squares for a component of the model by its degrees of freedom. The mean square for between-groups is calculated as

$$MS_B = \frac{SS_B}{df_B}.$$

In this example, the sums of squares for between-groups was 723.47, and there were 2 degrees of freedom for between-groups, so

$$MS_B = \frac{723.47}{2} = 361.73.$$

The mean square within groups is

$$MS_W = \frac{SS_W}{df_W}.$$

In this example, the within-groups sum of squares was 462 with 27 degrees of freedom, so

$$MS_W = \frac{462}{27}$$
$$= 17.11.$$

The final step is to obtain the F-ratio, which is the ratio of the mean square for between to mean square for within.

$$F = \frac{MS_B}{MS_W}$$

In this example,

$$F = \frac{361.73}{17.11}$$
$$= 21.14.$$

■ ANOVA Table

Statisticians often refer to the analysis of variance as ANOVA. The results are typically summarized in an ANOVA table. An example is Table 8.3. The table has six columns. The first column is for the source of variation: between, within, and total. The next three columns are for the sums of the squares, the degrees of freedom, and the mean squares. For each of these columns, there is an entry for both the between- and within-groups value. The total row is used only for sums of squares and degrees of freedom. (It is not worthwhile to sum mean squares.) The mean square can be obtained from the preceding two columns. (Remember $MS = SS/df.$)

The last two columns are for the F-value and for the probability level. The F-value is obtained from the information in the MS column ($F = MS_B/MS_W$). To find the p or probability level we must use the F-table in Appendix 3. F-values are always reported with their degrees of freedom. This problem has 2/27 degrees of freedom, so we go down the column labeled "Degrees of Freedom

Table 8.3					
Analysis of Variance					
SOURCE	SS	df	MS	F	p
Between	723.47	2	361.73	21.14	<.01
Within	462.00	27	17.11		
TOTAL	1185.47	29			

for Denominator" until we reach 27. Then we go across the row to the second column, which is associated with 2 degrees of freedom for the numerator. Suppose that we had set the significance level for this problem at .01. According to the table, an F-value must exceed 5.49 with 2/27 degrees of freedom to be statistically significant. The obtained value of 21.13 clearly is larger than 5.49. Thus, we reject the null hypothesis of no difference between treatment groups and conclude that the observed differences between the three treatment groups are statistically significant.

■ Step-by-Step Calculation of the Analysis of Variance

The example in Table 8.2 shows that the analysis of variance can be calculated from the definitional formulas. However, there are more direct computational formulas that can be used. These formulas are shown in Box 8.3. There is also a computational example with step-by-step instructions in Box 8.4. You should work through the calculation of the analysis of variance. If you feel comfortable with the formulas given in Box 8.3, you may wish to try a problem without the step-by-step instruction. However, if you run into trouble you may find the step-by-step instructions in Box 8.4 helpful. One word of caution: this example was chosen to be somewhat difficult since it includes negative scores; it shows how all values in the analysis of variance table will be positive even though many of the scores are themselves negative.

The Multiple-Comparisons Problem

Using the one-way analysis of variance, it is possible to determine whether there are significant differences among three groups. The null hypothesis is that all groups are sampled from the same population. When the null hypothesis is rejected, we conclude that observed differences among group means should not be attributed to sampling error. However, the question of which specific means differ from one another is unanswered. Was it group 1 that differed from groups 2 and 3? Or was it group 3 that differed from groups 1 and 2? One solution is to perform multiple t-tests. In our example with three groups, three possible t-tests could be performed (group 1 vs. group 2, group 1 vs. group 3, group 2 vs. group 3). As more groups are evaluated in the one-way analysis of variance, greater numbers of t-tests are required. The number of possible comparisons for J-groups is defined as

$$\frac{J(J-1)}{2}.$$

(8.1)

● BOX 8.3

Computational Formulas for One-Way ANOVA

Table 8.2 defined the components of the analysis of variance. However, the formulas given in that table are often cumbersome to use in calculation. This box gives mathematically equivalent computational formulas.

Null Hypothesis

H_0: $\mu_1 = \mu_2 = \mu_3 = \cdots = \mu_J$ (for group 1, 2, 3, ... J)

Alternate Hypothesis

H_1: $\mu_1 \neq \mu_2 \neq \mu_3 \cdots \neq \mu_J$

Assumptions:
1. Scores drawn from normally distributed populations.
2. Cases selected from populations randomly and independently.
3. Variances of populations from which cases are drawn are equal.

Formulas

Source	SS	df	MS	F	p
Between-groups	$\sum\left[\dfrac{(\sum X_{.j})^2}{n_j}\right] - \dfrac{(\sum X_{..})^2}{N}$	$J - 1$	SS_B/df_B	MS_B/MS_w	determined from Appendix 3 for appropriate df
Within-groups	$\sum\left[(\sum X_{.j}^2)\right] - \dfrac{(\sum X_{.j})^2}{n_j}$ or Total$_{SS}$ − Between$_{SS}$	$N - J$	SS_w/df_w		
	Total $\sum X_{ij}^2 - \dfrac{(\sum X_{ij})^2}{N}$	$N - 1$			

Computations

Source	SS	df	MS	F	p
Between-groups	$\left[\dfrac{(-120)^2}{10}\right.$ $\left. + \dfrac{(-26)^2}{10} + \dfrac{(-8)^2}{10}\right]$ $- \dfrac{(-154)^2}{30} = 723.47$	$3 - 2 = 1$	$\dfrac{723.47}{2} = 361.73$	$\dfrac{361.73}{17.11} = 21.14$	.01
Within-groups	$1185.47 - 723.46 = 462$	$30 - 3 = 27$	$\dfrac{462.01}{27} = 17.11$		
Total	$1976 - \dfrac{(154)^2}{30} = 1185.47$	$30 - 1 = 29$			

Calculation of One-Way Analysis of Variance for Data in Table 8.1

GROUP 1			GROUP 2			GROUP 3		
S	$X_{.1}$	$X_{.1}^2$	S	$X_{.2}$	$X_{.2}^2$	S	$X_{.3}$	$X_{.3}^2$
X_{11}	-6	36	X_{12}	0	0	X_{13}	2	4
X_{21}	-8	64	X_{22}	-2	4	X_{23}	-4	16
X_{31}	-21	441	X_{32}	-1	1	X_{33}	-6	36
X_{41}	-17	289	X_{42}	3	9	X_{43}	3	9
X_{51}	-16	256	X_{52}	-6	36	X_{53}	1	1
X_{61}	-14	196	X_{62}	-1	1	X_{63}	4	16
X_{71}	-11	121	X_{72}	-4	16	X_{73}	-5	25
X_{81}	-9	81	X_{82}	-1	1	X_{83}	-2	4
X_{91}	-8	64	X_{92}	-2	4	X_{93}	0	0
X_{101}	-10	100	X_{102}	-12	144	X_{103}	-1	1
$\Sigma X_{.1}$	-120		$\Sigma X_{.2}$	-26		$\Sigma X_{.3}$	-8	
$\Sigma X_{.1}^2$		1648	$\Sigma X_{.2}^2$		216	$\Sigma X_{.3}^2$		112
$N_1 = 10$			$N_2 = 10$			$N_3 = 10$		

STEPS

1. Find

$$\Sigma X_{..} = \Sigma X_{.1} + \Sigma X_{.2} + \cdots + \Sigma X_{.j}$$
$$= (-120) + (-26) + (-8)$$
$$= -154$$

2. Obtain

$$\Sigma X_{ij}^2 = \Sigma X_{.1}^2 + \Sigma X_{.2}^2 + \cdots + \Sigma X_{.j}^2$$
$$= 1648 + 216 + 112$$
$$= 1976$$

The notation X_{ij} refers to individual cases. The individual scores are squared and summed for each group, and then summed across groups.

3. Find $\Sigma X_{..}/N$ by dividing the result of Step 1 by N, the *total* number of cases.

$$\frac{\Sigma X_{..}}{N} = \frac{\text{Step 1}}{N} = \frac{-154}{30}$$
$$= -5.13$$

4. Get $(\Sigma X_{.1})^2/n_1$ which is the sum of the squares for group 1.

$$= \frac{(-120)^2}{10} = 1440$$

5. Get $(\Sigma X_{.2})^2/n_2$, the sum of squares for group 2.

$$= \frac{(-26)^2}{10} = 67.6$$

6. Find $(\Sigma X_{.3})^2/n_3$ which is the sum of squares for group 3

$$= \frac{(-8)^2}{10} = 6.4$$

(Repeat this step for each additional group if there are more than 3 groups in the analysis.)

(*continued*)

• BOX 8.4 (*continued*)

Calculation of One-Way Analysis of Variance for Data in Table 8.1

7. Calculate the total square sum of scores divided by n_j.

$$\frac{\Sigma(\Sigma X_{.j})^2}{n_j}$$

This is equal to the sum of the results from Steps 4, 5, and 6 (for the three group experiment).

$$= 1440 + 67.6 + 6.4$$
$$= 1514$$

8. Calculate $(\Sigma X_{..})^2/N$ by squaring Step 1 and dividing by the total N.

$$= \frac{(-154)^2}{30} = 790.53$$

9. You are now ready to begin filling in the ANOVA Summary Table. First, get the sum of squares for between-groups by subtracting Step 8 from Step 7.

$$= 1514 - 790.53$$
$$= 723.47$$

10. The within-groups sum of squares is Step 2 minus Step 7.

$$= 1976 - 1514$$
$$= 462$$

11. The total sum of squares is Step 2 minus Step 8.

$$= 1976 - 790.53$$
$$= 1185.47$$

12. The degrees of freedom for between-groups are the number of groups minus 1.

$$J - 1 = 3 - 1$$
$$= 2$$

13. The degrees of freedom for within-groups are the total number of subjects minus the number of groups.

$$N - J = 30 - 3$$
$$= 27$$

14. The total degrees of freedom is the total number of subjects minus 1.

$$N - 1 = 30 - 1$$
$$= 29$$

15. The mean square for between-groups is the sum of squares between groups divided by its degrees of freedom or Step 9 divided by Step 12.

$$\frac{\text{Step 9}}{\text{Step 12}} = \frac{723.47}{2}$$
$$= 361.73$$

16. The mean square for within-groups is the sum of squares within groups divided by the degrees of freedom within or Step 10 divided by Step 13.

$$\frac{\text{Step 10}}{\text{Step 13}} = \frac{462}{27}$$
$$= 17.11$$

17. The F-value is the ratio of mean squares between to mean squares within.

$$\frac{\text{Step 15}}{\text{Step 13}} = F_{(2/27)}$$
$$= \frac{361.73}{17.11}$$
$$= 21.14$$

For example, with five groups the number of comparisons required would be

$$\frac{5(5-1)}{2} = 10.$$

As we noted earlier there is a serious problem with making a large number of paired comparisons. As the number of comparisons increases, so does the probability that we will falsely reject the null hypothesis. In recognition of this problem, statisticians have created special tests to evaluate differences between group means after the null hypothesis has been rejected in the analysis of variance. Some of these methods are called *a priori* because they are planned before the analysis of variance is used. Other procedures are called **post hoc** because the experimenter had not planned exactly which means would be compared in advance. These statistical methods, which are based on different assumptions, produce different results. There are now many different methods that can be used to compare individual means in the analysis of variance. However, a review of all of these methods is beyond the scope of this text and is covered in detail in more advanced texts (Kirk, 1982).

There is one procedure that can be used relatively easily. This is known as the Fisher Least Significant Difference (LSD) Test. The Fisher LSD test is a post hoc test that is easy to use and appropriate in a variety of circumstances (Cramer and Swanson, 1973). The Fisher LSD test should only be used after the null hypothesis in the one-way analysis of variance have been rejected. The test is simply a t-test that uses the mean square within groups as an estimate of variance. The test is calculated as follows.

$$t = \frac{\bar{X}_1 - \bar{X}_2}{\sqrt{MS_W\left(\dfrac{1}{n_1} + \dfrac{1}{n_2}\right)}} \tag{7.7}$$

It is evaluated for statistical significance according to the t-table with

$$df = N - J.$$

In the example of the weight loss experiment, consider the difference between the behavior therapy group (group 1) and the attention control group (group 2). The means for these two groups were -12 and -2.6, respectively. So the LSD test would be

$$t = \frac{-12 - (-2.6)}{\sqrt{17.11\left(\dfrac{1}{10} + \dfrac{1}{10}\right)}} = \frac{-9.4}{1.85} = -5.08.$$

The degrees of freedom are the same as for MS_W. They are

$$N - J = 30 - 3 = 27$$

according to Appendix 2. The difference between these two groups is statistically significant beyond the .01 level.

Now we check the difference between groups 1 and 3. The denominator of the LSD equation will be the same (since N_1 and N_2, and MS within are the same). So the t-value will be

$$t = \frac{-12 - (-.8)}{1.85} = \frac{-11.2}{1.85} = -6.05.$$

With 27 degrees of freedom this difference is also statistically significant beyond the .01 level.

Finally, we will evaluate the difference between groups 2 and 3. Again, the denominator of the LSD equation will be the same. So

$$t = \frac{-2.6 - (-.8)}{1.85} = \frac{-1.8}{1.85} = -.97.$$

With 27 degrees of freedom, this difference is not statistically significant. Thus, we conclude that group 1 differed significantly from both groups 2 and 3, but that groups 2 and 3 did not differ significantly from one another.

Summary

The one-way analysis of variance can be used to compare three or more groups. By analogy, the one-way analysis of variance might be considered an extension of the t-test when more than two groups are used in an experimental study.

In the t-test the numerator of the test statistic is the difference between the means. In the analysis of variance, the numerator is the variance across treatment means. The logic of the analysis of variance requires two independent estimates of variance. One is obtained from the variance between group means while the other is an independent estimate obtained from variance within groups. The ratio of these two variances is used to obtain the F-statistic.

Key Terms

1. **Partitioning**   The division of the total variance into independent components. In the one-way analysis of variance, the total variance is separated into between-groups variation and within-groups variation.
2. **Between-groups Variance**   The variation of cell means around the grand mean.
3. **Within-groups Variance**   The variation of individual cases around their cell means.

4. **Mean Square Between**   An estimate of the between-groups variance.
5. **Mean Square Within**   An estimate of within-cell variance.
6. **F-Ratio**   The ratio of mean square between/mean square within.
7. **Multiple Comparison Problem**   When more than two groups are compared using *t*-tests, the probability of rejecting the null hypothesis falsely (a type I error) increases.
8. **Post Hoc Test**   A special test used to evaluate the differences between pairs of means after the null hypothesis has been rejected in the one-way analysis of variance.

Exercises

1. Suppose you were comparing the differences between the means of 5 groups. What difficulties will you encounter if you use *t*-tests? How many *t*-tests would be required?

2. Three junior high schools participated in an experiment on smoking prevention. School #1 received a special social learning intervention, School #2 received a series of pamphlets about the evils of smoking, and School #3 received no treatment at all. At the end of the 7th grade year, students were asked to rate their attitude toward smoking on a 10 point scale, where 1 is strongly against smoking and 10 is favorable toward smoking. Data from 10 students in each of the 3 schools are presented below:

STUDENT	SCHOOL #1	SCHOOL #2	SCHOOL #3
1	2	1	6
2	3	3	5
3	1	4	4
4	2	2	2
5	3	1	6
6	4	5	4
7	1	5	7
8	1	4	1
9	5	5	6
10	1	6	7

Use the one-way analysis of variance to compare the differences between the three schools. State the null hypothesis and evaluate it.

3. Is the Fisher LSD Test appropriate for the data in Exercise 2? If so, apply it and interpret the results.

9

The Two-Way Analysis of Variance

If there is one thing we have learned about behavioral phenomena, it is that they are complex. Organisms respond to specific stimulation under some circumstances but not others. For example, television violence may stimulate aggressive behavior when the viewers are angry, but have no effect if the viewers are not angry (Kaplan and Singer, 1976). To understand this it is necessary to manipulate two variables in the same experiment: television violence (by showing a violent film or a neutral film) and subject anger level (manipulated by making subjects angry or not angry).

In Chapter 8, methods for the simple, or one-way, analysis of variance were presented. These methods are used to evaluate the effect of one independent variable with two or more levels upon a dependent variable. The one-way analysis of variance can be used to evaluate the effects of television content upon aggression. For example, an experiment might be conducted in which three films are shown: a violent film, an equally arousing but nonviolent sports film, and a nonarousing control film. The dependent measure might be level of observed aggression following the film. Early experiments of this nature rarely demonstrated that there was any impact of televised violence. However, later experiments showed that television violence stimulates aggression when the subjects are angry prior to watching the television film. To understand the complexity of the situation, it was necessary to do experiments in which television violence and level of viewer anger were manipulated simultaneously. There are three separate questions addressed in such studies. First, there is the question of the impact of television violence upon aggression. Second, there is the effect of anger provocation upon aggressive behavior. And third, there is the issue of unique combinations of television violence and anger. In other words, the third question asks whether unique combinations of being angry and watching television violence result in increased aggressive behavior. The one-way analysis of variance is incapable of answering these three different questions. The two-way ANOVA provides a method that allows independent estimates of each of these experimental effects.

Another example of the need for a more complex analysis of variance model comes from studies of judgment bias. Some studies have shown that judges will rate the quality of an essay as superior when they are led to believe that it was written by a very attractive woman in comparison to evaluations of the same essay when the judges thought that the author was an unattractive woman. In this experiment, exactly the same essay was given to student judges who were asked to evaluate its quality and the talent of the author. The author was always described in exactly the same way, yet, for some of the subjects, a picture of a very attractive woman (supposedly the author) was attached to the essay. For another group, a picture of a very unattractive woman was attached. Even though the essay for the two groups was identical, the attractive woman was judged as being more talented. This result led some social psychologists to speculate that "beauty is talent" (Landy and Sigall, 1974).

Later, some social psychologists began to raise questions about the study. One problem was that all of the subjects in the experiment were male. Could it be that female judges would have reacted the same way to attractive and unattractive female authors? The experiment was replicated with both male and female judges. The finding that men are biased by the attractiveness of the author was replicated. However, it was also found that female judges exhibited a slight bias *against* the attractive female author. Understanding the outcome requires that we attend to the complexity of the situation. If the experiment had not simultaneously considered the effect of author attractiveness and sex of judge, we would not have learned that it is the unique combination of these two variables that defines the bias.

This chapter presents methods for considering two variables within the same experiment. In particular, we focus on experiments that use **factorial designs.** The data analysis method for these experiments is typically two-way analysis of variance. Analysis of variance methods for very complex experiments are available. You will undoubtedly learn about these methods if you take advanced courses in experimental design. Or you can read about them in advanced textbooks (Kirk, 1982).

Factorial Designs

In a factorial design or factorial experiment, two or more independent variables are simultaneously manipulated. Each independent variable has a discrete number of levels. For example, the independent variable, sex, has two levels: male and female. In the factorial experiment, the independent variables are combined to form all possible combinations.

An example of a two-by-two factorial design is shown in Table 9.1. The example is taken from the study on the "attractiveness halo effect" discussed

Table 9.1

Data for attractiveness halo experiment

		VARIABLE B: AUTHOR	
		B_1 UNATTRACTIVE	B_2 ATTRACTIVE
VARIABLE A: SEX OF JUDGE	A_1 MALE	4	7
		3	8
		4	4
		4	6
		2	8
		5	6
		4	7
		5	4
		6	7
		5	6
		4	5
		3	7
	A_2 FEMALE	6	4
		5	4
		5	5
		7	6
		5	6
		8	5
		5	5
		7	7
		8	6
		6	4
		5	5
		6	4

above. Male and female judges evaluated an essay attributed to either an attractive or an unattractive female author. There are two independent variables in the experiment. The first independent variable, which we call Variable A, is sex of judge. There are two levels: male (A_1) and female, (A_2). The second independent variable, Variable B, is attractiveness of author. This also has two levels: unattractive (B_1) and attractive (B_2). The cells in the table represent unique combinations of the two variables. There are four such unique combinations. For example, Cell $A_1 B_1$, is the combination of level 1 of Variable A and level 1 of Variable B. In this case, it is when males judged an essay attributed to an unattractive woman. Another example is Cell $A_1 B_2$. In this case, male judges evaluate an essay attributed to an attractive woman.

Factorial designs can be very complex. In the example above, it would be easy to add an additional independent variable for sex of the supposed author of the essay. Then, the design would be $2 \times 2 \times 2$. This design would have eight unique combinations. You could also have more than two levels for any variable. A $4 \times 5 \times 3$ design would have 60 unique combinations to compare. However, more complex experimental designs are beyond the scope of this book. The remainder of this chapter focuses on experimental designs that have no more than two independent variables. Further, we focus on experiments in which the observations in each cell are independent of each other.

Main Effects and Interactions

The two-way analysis of variance provides evaluations of three different hypotheses. Two of these concern **main effects** and one is an **interaction.** In the experiment on the attractiveness halo effect, we might consider three separate questions:

1. Do male and female judges differ in their evaluations of the essays (independent of the attractiveness of the author)?
2. Does the attractiveness of the author influence judgment of the essay (independent of the sex of the judge)?
3. Is the effect of author attractiveness different for male and for female judges?

The first two questions concern main effects while the third question concerns an interaction.

■ Main Effects

A main effect of a variable is an experimental effect that is independent of the influence of other variables. For example, the main effect of sex of judge in the attractiveness halo effect study is the difference between male and female judges averaged over levels of the other variable—in this case, attractiveness of the essay author. In the two-way analysis of variance, there are two main effects. One is the effect of A averaged over levels of Variable B and the other is the effect of B averaged over levels of Variable A. In the attractiveness halo effect study, these are the effects of sex of judge independent of the attractiveness of the author, and the effect of attractiveness of the author independent of sex of the judge.

When main effects are evaluated, we consider the effect of one variable and ignore the effect of the other. In an experiment, it is possible to observe a significant effect of Variable A when the effect of Variable B is nonsignificant. It is also possible to find a significant effect of Variable B when Variable A is nonsignificant. It is also possible that both variables will be significant or both variables will not be significant. In other words, the effects of the independent variables are independent of one another. The analysis of variance model and factorial experimental design assures that the estimation of these effects is statistically independent.

■ Interactions

Interactions describe the joint effects of two variables. An interaction between two variables might be described as a difference in differences. We ask whether the difference between levels of Variable A is the same at each level of Variable B. For example, is the difference in bias attributable to the attractiveness of the author of an essay the same for male judges as it is for female judges? When the effect of Variable B is the same for each level of Variable A, we say that the model is **additive.** This means that the experimental effects can be described from a combination of Variable A plus Variable B. However, if the differences between levels of one variable are *contingent* upon levels of the other variable, we say that the effect is **multiplicative.** This means that the results cannot be described by the additive effects of the two independent variables.

The interpretation of the results of the two-way analysis of variance is often aided by graphs. Figures 9.1–9.4 show a series of hypothetical graphs from the 2-by-2 experiment on the attractiveness halo effect. These are just a few of the many possible outcomes from such an experiment. To obtain the points on the graph, means are obtained from each of the four cells of the experiment. Then, one of the independent variables is chosen as the ordinate of the graph. In this case, we have chosen the attractiveness of author variable. Above the label "attractive," the cell means for males rating the essay attributed to the attractive author and females rating the essay attributed to the attractive author are plotted. Similarly, above "unattractive," we have plotted the cell means for males and for females rating the essay attributed to the unattractive woman.

Figures 9.1–9.4 show four alternative outcomes of a two-factor experiment.

Figure 9.1

Main effect for sex of subject; main effect for attractiveness; no interaction.

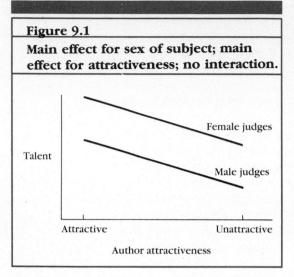

Figure 9.2

Main effect of attractiveness; no effect for sex of subject; no interaction.

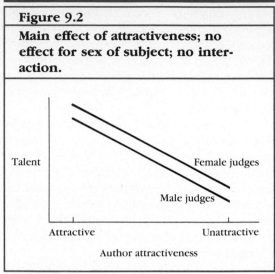

Figure 9.3

No main effects or interaction.

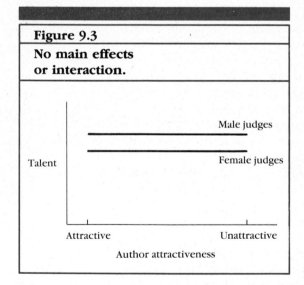

Figure 9.4

No main effects; significant interaction.

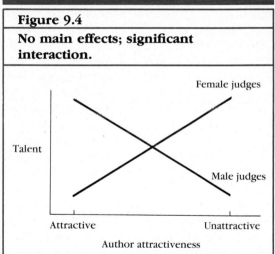

Figure 9.1 shows a main effect for both sex of judge and attractiveness of author, but no interaction. The main effect for sex of judge is apparent from the difference between the line for female judges and the line for male judges. Averaged over levels of attractiveness, there is a difference between the way males and females rate the pictures. The main effect for attractiveness is suggested because both male and female judges rate the work by the attractive au-

thor as superior to that of the unattractive author. In other words, averaged over sex of judge there is a difference between ratings of an essay attributed to an attractive or an unattractive author.

Figure 9.2 shows no effect of sex of judge, an effect of author attractiveness, and no interaction. Here the lines for the male and female judges are very close to one another and do not differ significantly. However, both male and female judges rate the essay attributed to the attractive woman as exhibiting more talent than the essay attributed to the unattractive woman. The differences between male and female judges are essentially the same for ratings of both the attractive and unattractive author. Therefore, there is no interaction. Figure 9.3 shows neither a main effect nor an interaction. Looking back at the first three figures in which there is no interaction, note that the lines in the graph are **parallel** in all cases. This is the defining characteristic of the additive model or absence of interaction.

An example of an interaction is shown in Fig. 9.4. When there is a significant interaction between the independent variables in the two-way analysis of variance, the lines on a graph will *not* be parallel. An interaction is a difference of differences. In this example, the differences between ratings of work attributed to an unattractive or an attractive woman are different for male raters than they are for female raters. Female raters rate the author as more talented if she is unattractive than if she is attractive. Conversely, male raters rate the author as more talented if she is attractive than if she is unattractive. Another way of describing this result is to say that the effect of author attractiveness is *contingent* upon the gender of the raters.

The Theory of the Two-Way Analysis of Variance

The logic underlying the two-way analysis of variance is an extension of that for the one-way analysis of variance. By way of review recall that the one-way analysis of variance involved the partitioning of the total sums of squares into two separate components. One component is the between-groups sums of squares, while the other is the within-groups sums of squares. The between-groups sums of squares represents the variation of group means around the grand mean. When divided by the degrees of freedom for between-groups, the between-groups sums of squares becomes the between-groups mean square. The mean square is an estimate of variance of group means. The within-groups sums of squares is an estimate of the variation of individual scores around their group mean. The assumption is that this variation is approximately equal in each group. Thus, the average variation of individual scores around their group mean is an estimate of the degree to which scores vary on the particular outcome measure. In other words, this is an estimate of measurement error. Dividing the within-groups sums of squares by the degrees of freedom for within-groups gives the within-group mean square.

The between-groups and within-groups mean squares are each independent estimates of variance. If the group means do not differ, the two estimates are expected to be approximately the same. However, if the group means are different, the variability across means (*MS*-between) will be greater than the variability within the groups. Thus, the ratio of between- to within-groups mean squares will be greater than 1. This ratio is called the *F*-ratio and tables are used to determine if the *F*-ratio is *significantly* greater than 1.

The logic of the two-way analysis of variance is very similar. Instead of partitioning the sums of squares into two pieces, the two-way analysis of variance partitions the sums of squares into four independent estimates. The partitioning of the sums of squares for the two-way analysis of variance is shown in Box 9.1. To understand this partitioning, we must consider the score for some individual. In the example, we call this individual Herman, and we can represent his score as

$$X_{ijk}.$$

This means that Herman is the ith person in the jth level of Factor A and the kth level of Factor B. Partitioning involves consideration of all sources of variation around the grand mean. There are four separate and independent sources in the two-way analysis variable.

1. Variation of the mean of Factor A around the grand mean.
2. Variation of the mean of Factor B around the grand mean.
3. Variation of unique combinations of $A \times B$ around the grand mean.
4. Variation of individual scores around their cell or group means.

Each of the first three sources is a between-groups factor. The fourth source is a within-groups factor. Analogous to the one-way analysis of variance, each of these sources of variation provides an independent estimate of variance. The fourth estimate, or the estimate derived from individual variations around their group mean, is an estimate of error variance. The logic of the analysis of variance requires that we divide each of the other between-groups sources of variation by this within-groups factor to obtain *F*-values. For example, to determine the effect of Factor A, we obtain an estimate of the variation of means of the two levels of Factor A around the grand mean. Then we divide this variance estimate by the within-groups variance—an estimate of error variance. The ratio of these two variances is the *F*-ratio for Factor A. If an *F*-ratio significantly exceeds 1.0, this suggests that the variation or difference between levels of Factor A exceeds what would be expected by chance. Box 9.1 provides a step-by-step description of the partitioning of the sums of squares.

• BOX 9.1

The Partitioning of the Sums of Squares in the Two-Way ANOVA

The logic of the two-way ANOVA is similar to that of the one-way case. However, in the two-way ANOVA the total variation can be partitioned into four sources. To understand the pieces, let's consider one individual. We will call him Herman. Each score, including Herman's, is represented as

X_{ijk},

which means the ith person in the jth level of Factor A and the kth level of Factor B.

Suppose that Herman is the third person in the group and received level 1 of Factor A and level 2 of Factor B. Then Herman's score is represented as X_{312}. His score is 6.

		FACTOR B		
		LEVEL 1	LEVEL 2	
		Score	**Score**	
		$X_{111} = 4$	$X_{112} = 9$	
		$X_{211} = 6$	$X_{212} = 5$	
	LEVEL 1	$X_{311} = 3$	$X_{312} = 6$ ..Herman	$\bar{X}_{.1.} = 5.1$
		$X_{411} = 2$	$X_{412} = 7$	
		$X_{511} = 1$	$X_{512} = 8$	
		$\bar{X}_{.11} = 3.2$	$\bar{X}_{.12} = 7.0$	
FACTOR A		$X_{121} = 7$	$X_{122} = 1$	
		$X_{221} = 9$	$X_{222} = 2$	
	LEVEL 2	$X_{321} = 8$	$X_{322} = 1$	$\bar{X}_{.2.} = 4.6$
		$X_{421} = 7$	$X_{422} = 3$	
		$X_{521} = 6$	$X_{522} = 2$	
		$\bar{X}_{.21} = 7.4$	$\bar{X}_{.22} = 1.8$	
		$\bar{X}_{..1} = 5.3$	$\bar{X}_{..2} = 4.4$	$\bar{X}_{...} = 4.85$

Suppose Sophia is the fifth person in the group receiving level 2 of Factor A and level 1 of Factor B. Can you find her score? (*Answer:* It is the last entry in the lower left quadrant, X_{521}.)

The total variation can be represented as

$X_{ijk} - \bar{X}_{...}$,

using the example of Herman or the variation of each individual around the grand mean. The grand mean ($\bar{X}_{...}$) is 4.85. Herman's score is 6, so the variation of Herman's score from the grand mean is

$6 - 4.85 = 1.15.$ (*continued*)

The total variation is composed of several components. First there is the variation of the mean level of Factor A around the grand mean.

$$\bar{X}_{.j.} - \bar{X}_{...}$$

Herman is in level 1 of Factor A ($\bar{X}_{.1.} = 5.1$), so the variation of his Factor A mean from the grand mean is

$5.1 - 4.85 = .25.$

Next there is the variation of mean level of Factor B around the grand mean.

$$\bar{X}_{..k} - \bar{X}_{...}$$

Herman is in level 2 of Factor B($\bar{X}_{..2} = 4.4$), so this variation is

$4.4 - 4.85 = -.45.$

Then we must consider the unique cell with which Herman is associated, after adjusting for the effects of his association with Factor A and Factor B:

$$\bar{X}_{.jk} - \bar{X}_{.j.} - \bar{X}_{..k} + \bar{X}_{...}.$$

For Herman this quantity is

$7 - 5.1 - 4.4 + 4.85 = 2.35.$

Finally, there is the variation of a single case from its own cell mean:

$$X_{ijk} - \bar{X}_{.jk}.$$

For Herman, this is

$6 - 7 = -1.$

The Computation of the Two-by-Two Analysis of Variance

The computation of the two-by-two analysis of variance is illustrated in Box 9.2. The example uses the data from the attractiveness halo effect study. To summarize the design of the experiment, each participant in the experiment read an essay and rated the talent of the essay's author. Attached to each essay was a

Now we can check to be sure that the variation of an individual score around the grand mean can be broken into the components.

Herman's variation around the grand mean (1.15) should equal his Factor-A deviation (.25) plus his Factor-B deviation ($-.45$) plus his unique cell deviation (2.35) plus his personal deviation from his cell mean (-1). So we must check to see if $1.15 = (.25) + (-.45) + (2.35) + (-1)$. Indeed it does! Now that the total deviation is partitioned for an individual, we can sum the deviations. However, as with the deviations around the mean (see Chapter 3), the sum of the deviations will always be zero. Therefore, we must sum the *squared* deviations. The following table shows the components and their computational formulas.

COMPONENT	DEVIATION	SUM OF SQUARED DEVIATION	COMPUTATIONAL FORMULA
Total	$X_{ijk} - \bar{X}...$	$\Sigma(X_{ijk} - \bar{X}...)^2$	$\Sigma X_{ijk}^2 - \dfrac{(\Sigma X...)^2}{N}$
Factor A	$\bar{X}_{.j.} - \bar{X}...$	$\Sigma(X_{.j.} - \bar{X}...)^2$	$\dfrac{(\Sigma X_{.j.})^2}{n_j} - \dfrac{(\Sigma X...)^2}{N}$
Factor B	$\bar{X}_{..k} - \bar{X}...$	$\Sigma(X_{.k} - \bar{X}..)^2$	$\dfrac{(\Sigma X_{..k})^2}{n_k} - \dfrac{(\Sigma X...)^2}{N_k}$
Interaction	$\bar{X}_{.jk}\ \ \bar{X}_{.j.}$ $- \bar{X}_{..k} + \bar{X}...$	$\Sigma(\bar{X}_{jk} - \bar{X}_j.$ $- \bar{X}_{.k}. + \bar{X}...)^2$	$\dfrac{(\Sigma X_{.jk})^2}{n_{jk}} - \dfrac{(\Sigma X_{.j.})^2}{n_j}$ $- \dfrac{(\Sigma X_{..k})^2}{n_k} - \dfrac{(\Sigma X...)^2}{N}$

Where n_j = number of cases within level of j
$\quad n_k$ = number of cases within level of k
$\quad n_{jk}$ = number of cases within a combination of j and k

picture of the supposed author. The attractiveness of the female author was systematically varied. Half of the subjects rated the talent of an author who was shown to be very attractive. Others of the subjects rated essays of an author who was shown to be unattractive. This is Factor B, in the experimental design. The other factor is for sex of subject. Half of the subjects were male and half of the subjects were female. Sex of subject, then, provides the two levels of Factor A.

The data for the experiment are shown in Table 9.1. There were 12 subjects in each of the four conditions of the experiment. For example, 12 men rated the talent of an unattractive author and 12 different males rated the talent

● BOX 9.2

Calculation of the Two-Way Analysis of Variance (Data from Table 9.1)

STEPS

1. Find the $\Sigma X_{.jk}$ for each individual cell in the experiment. This sum would be labeled $\Sigma X_{.11}$, $\Sigma X_{.21}$, $\Sigma X_{.12}$, and $\Sigma X_{.22}$.

   $\Sigma X_{.11} = 4 + 3 + 4 + 4 + \cdots + 3 = 49$
   $\Sigma X_{.21} = 6 + 5 + 5 + 7 + \cdots + 6 = 73$
   $\Sigma X_{.12} = 7 + 8 + 4 + 6 + \cdots + 7 = 75$
   $\Sigma X_{.22} = 4 + 4 + 5 + 6 + \cdots + 4 = 61$

2. Count the number of observations in each cell. In the example, there are 12 in each cell: $n_{.11} = 12$, $n_{.12} = 12$, $n_{.21} = 12$, $n_{.22} = 12$.

3. Obtain the marginal sums for Factor A. (The marginal sums are the sums in the margins of the table.)

   $\Sigma X_{.1.} = \Sigma X_{.11} + \Sigma X_{.12} = 49 + 75 = 124$
   $\Sigma X_{.2.} = \Sigma X_{.21} + \Sigma X_{.22} = 73 + 61 = 134$

4. Find the marginal sums for Factor B.

   $\Sigma X_{..1} = \Sigma X_{.11} + \Sigma X_{.21} = 49 + 73 = 122$
   $\Sigma X_{..2} = \Sigma X_{.12} + \Sigma X_{.22}$
   $\qquad = 75 + 61 = 136$

5. Find the number of cases in each row and column total.

   For Factor A: $\quad N_{a1} = 12 + 12 = 24$
   $\qquad\qquad\qquad\quad N_{a2} = 12 + 12 = 24$
   $\qquad\qquad\qquad\quad N_{b1} = 12 + 12 = 24$
   $\qquad\qquad\qquad\quad N_{b2} = 12 + 12 = 24$

6. Sum the marginal totals for *either* factor to obtain the grand sum.

   $\Sigma X_{...} = \Sigma X_{..1} + \Sigma X_{..2}$ or $\Sigma X_{.1.} + \Sigma X_{.2.}$
   $\qquad = 122 + 136$
   $\qquad = 258$
   or
   $\qquad\qquad = 124 + 134 = 258$

   Use the result from one factor as a check against the other.

7. Get N or the total number of observations.

   $N = 48$

8. Obtain $(\Sigma X_{...})^2/N$ using the results of Steps 6 and 7.

$$= 258^2/48$$
$$= 1386.75$$

9. Find $\Sigma X_{...}^2$ by squaring each observation and summing the results.

$$4^2 + 3^2 + 4^2 + \cdots + 6^2 + 5^2 + 5^2 + \cdots + 7^2 + 8^2 + 4^2 + \cdots$$
$$+ 4^2 + 4^2 + 5^2 + \cdots + 4^2 = 1482$$

10. Find SS-total by subtracting the results of Step 8 from Step 9.

$$1482 - 1386.78 = 95.22$$

Note: This should always be a positive number!
If it is negative, check Steps 8 and 9.

11. Obtain SS_A through the following substeps.

a) Obtain

$$\frac{(\Sigma X_{.1.})^2}{N_{a1}} + \frac{(\Sigma X_{.2.})^2}{N_{a2}}$$

using results from Steps 3 and 5:

$$= \frac{(124)^2}{24} + \frac{(134)^2}{24}$$
$$= 640.67 + 748.17$$
$$= 1388.84.$$

b) Get SS_A by subtracting the results of Step 8 from Step 11(a):

$$1388.83 - 1386.78 = 2.06.$$

12. Obtain SS_B through the following substeps:

a) Find

$$\frac{(\Sigma X_{..1})^2}{N_{b1}} + \frac{(\Sigma X_{..2})^2}{N_{b2}}$$

(continued)

using results from Steps 4 and 5:

$$= \frac{(122)^2}{24} + \frac{(136)^2}{24}$$

$$= 620.17 + 770.67$$

$$= 1390.84.$$

b) Calculate SS_B by subtracting the results of Step 8 from Step 12(a):

$$= 1390.84 - 1386.75 = 4.09.$$

13. Determine SS_{AB} through the following substeps:

   a) Sum the *squared* cell totals

$$= (\Sigma X_{.11})^2 + (\Sigma X_{.12})^2 + (\Sigma X_{.21})^2 + (\Sigma X_{.22})^2$$

   The totals were obtained in Step 1.

$$= 49^2 + 73^2 + 75^2 + 61^2$$

$$= 17076$$

   b) Divide the results of Step 13(a) by the number of observations in *each cell.*

$$\frac{17076}{12} = 1423$$

   c) SS_{AB} is obtained by subtracting Steps 8, 12(b), and 11(b) from Step 13(b):

$$SS_{AB} = 1423 - 1386.75 - 4.09 - 2.06 = 30.10.$$

14. Find SS-error $= SS$-total (Step 10) $- SS_A$ (Step 11b) $- SS_B$ (Step 12b) $- SS_{AB}$ (Step 13c)

$$= 95.22 - 4.09 - 2.08 - 30.10$$

$$= 58.97$$

15. Enter the Sums of Squares in an ANOVA table. The table has columns for Source, SS, df, MS, and F. The rows are A, B, $A \times B$, within, and total.

Source	SS	df	MS	F
A	2.06			
B	4.09			
$A \times B$	30.10			
Within	58.97			
Total	95.22			

16. Find the degrees of freedom for the different components as follows.

Component	df	Description	Result
A	$J - 1$	Number of levels of factor $A - 1$	$2 - 1 = 1$
B	$K - 1$	Number of levels of factor $B - 1$	$2 - 1 = 1$
$A \times B$	$(J - 1)(K - 1)$	The product of df for A and B	$(1)(1) = 1$
Within	$N - JK$	Total number of subjects minus the product of the levels of A times the levels of B	$48 - (2 \times 2) = 44$
Total $N - 1$		Total number of subjects minus 1	$48 - 1 = 47$

17. Enter the degrees of freedom for each component into the ANOVA Summary Table. Now divide each SS by the df in the same row. The result is the Mean Square or MS (no need to do this for the total row).

Source	SS	df	MS	F
A	2.06	1	2.06	
B	4.09	1	4.09	
$A \times B$	30.10	1	30.10	
Within	58.97	44	1.34	
Total	95.22			

The MS for A was found by dividing the SS for A by the df in that same row:

$$\frac{2.06}{1} = 2.06.$$

In the 2×2 ANOVA, the MS for A, B, and $A \times B$ will always be the same as the SS for these factors because the SS is always divided by 1 (the df for each factor in the 2×2 design).

18. Calculate the F for A, B, and $A \times B$ by dividing each by the MS for error.

$$F_A = \frac{MS_A}{MS_W} = \frac{2.06}{1.34} = 1.53$$

$$F_B = \frac{MS_B}{MS_W} = \frac{4.06}{1.34} = 3.05$$

$$F_A \times B = \frac{MS_{A \times B}}{MS_W} = \frac{30.10}{1.34} = 22.46$$

(continued)

● BOX 9.2 (*continued*)

Source	SS	df	MS	F
A (Sex of Rater)	2.06	1	2.06	1.53
B (Attractiveness of Target)	4.09	1	4.06	3.05
A × B	30.10	1	30.10	22.46*
Within	58.96	44	1.34	
Total	95.22	47		

*$p < .01$

19. Go to Appendix 3 to find the criterion values for significant F-ratios. Look up the value corresponding to the appropriate *df,* for both numerator and denominator. For this example, we need the F-values for 1/44 degrees of freedom. According to the appendix, a significant F at the .05 level must exceed 4.06; at the .01 level it must exceed 7.24. In our example, only the F for the interaction is significant, and it exceeds the .01 level.

20. To interpret a significant interaction, draw a two-way graph. Place levels of one independent variable on the *x*-axis, and use the *y*-axis for the dependent variable. Plot points for each level of the other independent variable.

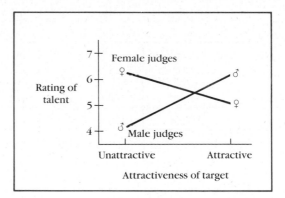

Significant interactions are reflected by nonparallel lines on the graph.

of an attractive author. Similarly, 12 females rated the talent of an unattractive author while 12 different females rated the talent of an attractive author. Although it is relatively easy to calculate the two-way analysis of variance using computer programs, it will be worth your while to make the computations by hand a few times. This will give you a better feel for the analysis of variance and how it works. A step-by-step description of the calculations is shown in Box 9.2.

Interpretation of the *F*-Ratio

To determine if an *F*-ratio is statistically significant, we must use Appendix 3. This is the same table that is used to interpret the results of the one-way analysis of variance described in Chapter 8. The *F*-ratio is associated with two different degrees of freedom, one for the numerator and the other for the denominator. Appendix 3 is arranged such that the degrees of freedom associated with the numerator are shown as columns in the table and the degrees of freedom for the denominator are shown as rows. Each combination of degrees of freedom for numerator and denominator has two different values in the table: one for the .05 significance level and the other for the .01 significance level. For example, look at the *F*-values required for one degree of freedom for the numerator and one for the denominator in the upper left corner of the table. At the .01 significance level, we would require an *F*-ratio of 4052.

Now let's try a value more similar to that in the usual experiment we would be conducting. Let's take 1 degree of freedom for the numerator and 28 degrees of freedom for the denominator. To find the criterion value, we look down the first column until we come to the row adjacent to 28 degrees of freedom for the denominator (or the 28th row). The table tells us that the .05 significance level is associated with F-values equal to or greater than 4.20. For the same numbers of degrees of freedom, the .01 significance level is associated with an F-value of 7.64 or greater.

Suppose that we have conducted an experiment, and we have obtained an F-ratio with 1/21 degrees of freedom. The F-value is 3.75. Is it statistically significant? To determine this, we go through the table and find the column associated with 1 degree of freedom for the numerator and the row associated with 21 degrees of freedom for the denominator. At the .05 significance level, we need 4.32 to obtain a significant effect. The observed F-value is less than 4.32, so we conclude that the difference between group means is not statistically significant. Let's try another hypothetical experiment in which we obtained an F-value of 9.23 with 1/60 degrees of freedom. Checking the table for the column associated with 1 degree of freedom and the row associated with 60 degrees of freedom, we find that 7.08 defines the lower boundary of the .01 significance level. This suggests that our observed value is greater than what would be required at the .01 level. In other words, the observed difference is statistically significant at the .01 significance level.

Assumptions

Whenever we perform the analysis of variance, we make several assumptions. First we assume that the samples are drawn randomly and independently from a defined population. Second we assume that the variances of the population are homogenous, or approximately the same. This is usually referred to as the homogeneity of variance assumption. If there are four groups in the experiment, we assume that the variances within each of these groups are approximately equal. The third assumption is that the scores are normally distributed within the populations. For the procedures presented in this chapter, we also assume that the number of cases in each condition is the same.

If we cannot assume these to be true, it may be necessary to use another statistical model to analyze the data. Box (1954) demonstrated that the analysis of variance is robust with regard to most of these assumptions. (*Robust* means that a violation of these assumptions has relatively little effect on the results of the analysis.) Nevertheless, if these assumptions cannot be made, it is important to get the advice of an expert statistician. In Chapter 12, we review some methods that are used when we know that these assumptions will be violated.

Summary

The two-way analysis of variance is typically used for experiments employing factorial designs. In factorial designs or factorial experiments, two or more independent variables are simultaneously manipulated. This chapter is limited to the 2×2 ANOVA. Using the 2×2 ANOVA, the experimenter can evaluate three separate hypotheses. First, she can estimate the effect of Factor A (independent of Factor B). Second, she can evaluate the effect of Factor B (independent of Factor A). And finally, she can estimate the effect of unique combinations of levels of factors A and B.

The effects of the unique combinations of variables are called interactions. They suggest that differences between levels of one variable are contingent upon levels of the other variable. When there is a significant interaction, we say that the additive model does not describe the data and that a multiplicative model is required. The two-way analysis of variance is a method commonly in use in experimental sciences. Advanced courses in statistics often focus on more complex models for analyzing data from factorial experiments.

Key Terms

1. **Factorial Design**   An experimental design in which two or more independent variables are simultaneously manipulated. Each independent variable has a discrete number of levels.
2. **Main Effect**   An experimental effect in a factorial design that is independent of the influence of other variables.
3. **Interaction**   An experimental effect in the analysis of variance that represents unique combinations of levels of two or more variables.
4. **Additive Effect**   In a factorial experiment, the effect of each factor that is independent of the effect of any other factor.
5. **Multiplicative Effect**   An experimental effect in which the differences between levels of one variable are contingent upon levels of another variable.
6. **Parallel Lines**   To interpret the results of a two-way analysis of variance, cell means are plotted on a graph in which levels of one independent variable are points on the x-axis and a separate line is drawn for each level of the other independent variable. If the lines connecting the points are parallel, there is no interaction between the two independent variables.

Exercises

1. Calculate a 2 × 2 ANOVA for the following data table.

		B			
		B_1		B_2	
A	A_1	6 7 8 9	10 8 11 9	3 2 4 3	6 4 1 1
	A_2	7 6 3 1	6 7 4 2	12 10 9 8	15 10 9 12

2. A music teacher wanted to compare two types of rehearsal schedules for a group of intermediate and advanced piano players. One schedule required 4 one-half hour rehearsals each week. The other required 2 one hour rehearsals each week.

   Each student rehearsed the same piece by Bach. After 4 weeks, she had each student play the piece and she recorded the number of errors. The results are shown in the following table.

		STUDENTS			
		Intermediate		Advanced	
SCHEDULE	4 half-hour sessions	7 9 18 4 6	9 11 14 12 9	3 0 1 2 0	2 0 4 1 3
	2 one-hour sessions	16 18 21 9 16	17 13 14 12 9	6 4 5 1 2	7 5 2 4 6

What conclusions should the teacher draw from this experiment?

Other Statistical Techniques III

This book can cover only a limited number of the many procedures available to statisticians and scientists. In Parts I and II, methods for descriptive and for inferential statistics were presented. Chapter 10 presents methods for correlation and regression. These are used both to describe scores and to evaluate the association between variables. Correlational methods are used for both descriptive and inferential purposes. They are included because they have widespread use in many different sciences.

The application of correlational methods is common in the field of psychometrics. Psychometrics is the study of mental measurements. When studying mental measurement, special statistical procedures are required. Chapter 11 is devoted to these techniques.

Most of the methods used in inferential statistics require that certain assumptions be met. However, there are known situations when these assumptions are not tenable. Chapter 12 presents alternative methods that can be used when assumptions underlying common inferential statistical methods have been violated. The chapter also presents methods for comparing frequency distributions.

Many different statistical methods are presented in this book. When learning the methods one at a time, it is common to think of the procedures as unrelated to one another. Chapter 13 will help tie together many of the different statistical procedures that you have learned. In addition, this chapter provides some guidance for choosing statistical methods for different research problems.

10

Correlation and Regression*

A banner headline in an issue of the *National Enquirer* read "FOOD CAUSES MOST MARRIAGE PROBLEMS." The article inside the magazine talked about "Startling Results of Studies by Doctors and Marriage Counselors." Actually the headline was not based on any systematic study. Rather, it used the opinions of some physicians and marriage counselors who felt that high blood sugar is related to low energy level, which in turn causes marital unhappiness.

Unfortunately, the *National Enquirer* did not report enough data for us to make an evaluation of the hypothesis. However, we feel comfortable concluding that an association between diet and divorce has not been established. Before we are willing to accept the magazine's conclusion we must ask many questions. This chapter focuses on one of the many issues raised in the report—the level of association between variables.[†]

The *Enquirer* tells us that diet and unhappiness are associated, but it does not tell us to what extent. Is the association greater than we would expect by chance? Is it a strong or a weak association?

---

*Portions of this chapter are taken from R. M. Kaplan and D. P. Saccuzzo, *Psychological Testing: Principles, Applications, and Issues*. Monterey: Brooks/Cole, 1982. Reproduced by permission.

[†] There were many other problems with the *Enquirer* report. The observation was based on the clinical experiences of some health practitioners who found that many couples who came in for counseling had poor diets. One major oversight was that there was no control group of people who were not having marriage problems. We do not know from the study whether couples having problems have poor diets in greater proportions than people in general. Another problem is that neither diet nor marital happiness was measured in a systematic way. Thus we are left with subjective opinions about the levels of these variables. Finally, we do not know the direction of causation: does diet cause unhappiness or does unhappiness cause poor diet? Another possibility is that some other problem (such as stress) may cause both poor diet and unhappiness.

Lots of things seem to be related. For example, life stress is associated with heart disease, training is associated with good performance in athletics, overeating is associated with indigestion. People are always telling us that things are associated. For some events the association is obvious. For example, the angle of the sun in the sky and the time of day are associated in a very predictable way. This is because time was originally defined by the angle of the sun in the sky. Other associations are less obvious. The association between performing well on the Scholastic Aptitude Test and obtaining good grades in college is one example.

Sometimes we do not know whether these events are meaningfully associated with one another. If we do conclude that events are associated in some fundamental way, it is important to have a precise index of the degree. This chapter discusses statistical procedures that allow us to make precise estimates of the degree to which variables are associated. These methods are very important, and we will refer to them frequently in the remainder of this book. The indexes of association used most frequently in testing are correlation, regression, and multiple regression.

The Scatter Diagram

Before presenting the measures of association we look at visual displays of the relationship between variables. In Chapters 2 through 4 we concentrated on univariate distributions of scores. These involve only one variable—each person has only one score. This chapter considers statistical methods for studying bivariate distributions, which have two scores for each individual. For example, when we study the relations between test scores and classroom performance, we are dealing with a bivariate distribution. Each of several persons has a score on the test and a score for classroom performance. Averaged over individuals, we want to know whether these two variables are associated.

A scatter diagram is a picture of the relationship between two variables. An example of a scatter diagram is shown in Fig. 10.1. The axes in the figure represent scales for two variables. Values of X are shown on the horizontal axis and values of Y on the vertical axis. Each point on the scatter diagram shows where a particular individual scored on both X and Y. For example, one person had a score of 1 on X and a score of 2 on Y. This point is circled in the figure. It can be located if we find 1 on the X-axis and go straight up to the level of 2 on the Y-axis. Each point indicates the scores for X and Y for one individual. As you can see, the figure presents a lot of information. Each point represents the performance of one person who has been assessed on two measures.

The next sections present methods for summarizing the information in the scatter diagrams by finding the straight line that comes closest to more points

Figure 10.1

A scatter diagram. The circled point shows a person who had a score of 1 on X and 2 on Y (see text). The line expresses the regression that describes the best linear relationship between X and Y.

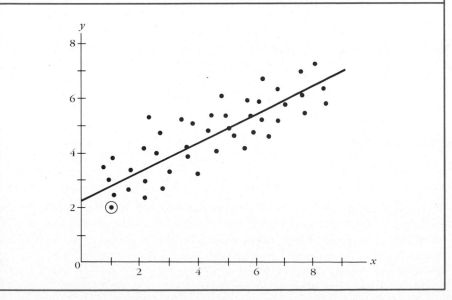

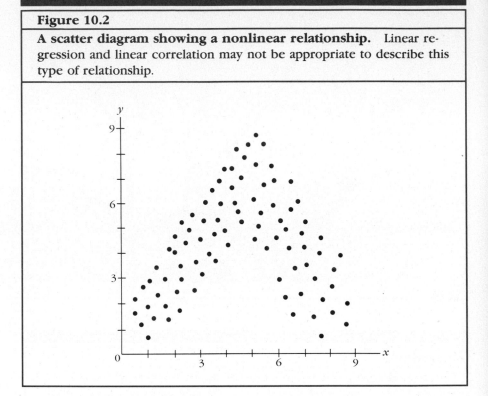

Figure 10.2

A scatter diagram showing a nonlinear relationship.   Linear regression and linear correlation may not be appropriate to describe this
type of relationship.

than any other. One of the reasons that it is important to examine the scatter
diagram is because the relationships between X and Y are not always best described by a straight line. For example, the relationship shown in Fig. 10.2 is
probably best described by a curved line. The methods of linear correlation or
linear regression, which are presented in this chapter, are not appropriate for
describing nonlinear relationships such as that illustrated in Fig. 10.2.

Correlation

In correlational analysis we ask if two variables *covary*. In other words, does Y
get larger whenever X gets larger? For example, does the patient feel dizzier
when we increase the dosage of the drug? Do people get more diseases when
they are under more stress? Correlational analysis is designed primarily to examine linear relationships between variables. Although we can use correla-

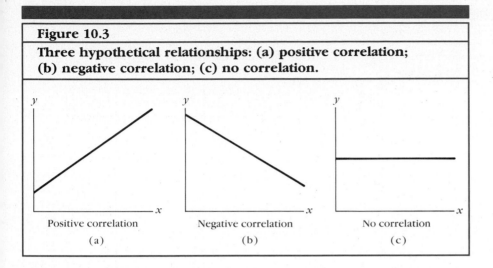

Figure 10.3

Three hypothetical relationships: (a) positive correlation; (b) negative correlation; (c) no correlation.

tional techniques to study nonlinear relationships, that is beyond the scope of this book.*

A **correlation coefficient** is a mathematical index that describes the direction and magnitude of the relationship. Figure 10.3 shows three different types of relationships between variables. The first section of the figure demonstrates a **positive correlation.** This means that high scores on Y are associated with high scores on X and low scores on Y correspond to low scores on X. The second section shows **negative correlation.** When there is a negative correlation, higher scores on Y are associated with lower scores on X, and lower scores on Y are associated with higher scores on X. This might describe the relationship between taking barbiturates and amount of activity. The higher the drug dosage, the less active patients will be. The third portion of Fig. 10.3 shows *no correlation* or a situation in which the variables are not related. Here scores on X do not give us information about scores on Y. An example of this sort of relationship would be the lack of correlation between shoe size and IQ.

There are a variety of ways to calculate a correlation coefficient. All involve pairs of observations: for each observation on one variable, there is an observation on one other variable for the same person.[†] Table 10.1 shows a set of observations on two variables arranged in ordered pairs. The observations are the number of games won and the average number of points scored by 28 teams in the National Football League during the 16 regular season games of

---

*Readers interested in methods for studying nonlinear relationships should review McNemar (1969).
[†]The pairs of scores do not always need to be for a person. They might also be for a group, an institution, a team, and so on.

Table 10.1

Average points scored per game and number of games won by ten teams during the 1984 NFL season

TEAM	X GAMES WON	Y POINTS	XY	X^2	Y^2
San Francisco	15	30	450	225	900
Denver	13	22	286	169	484
Washington	11	27	297	121	729
Pittsburgh	9	24	216	81	576
Cincinnati	8	21	168	64	441
NY Jets	7	21	147	49	441
Philadelphia	6	17	102	36	289
Detroit	4	18	72	16	324
Minnesota	3	17	51	9	289
Buffalo	2	16	32	4	256
	$\Sigma X = 78$	$\Sigma Y = 213$	$\Sigma XY = 1821$	$\Sigma X^2 = 774$	$\Sigma Y^2 = 4729$

the 1984 season. There are many ways to calculate a correlation coefficient, and these methods are all mathematically equivalent. Before we present methods for calculating the correlation coefficient, we first discuss regression. Regression is the basic method on which correlation is based.

Regression

■ The Regression Line

Correlation is used when we are interested in assessing the magnitude and direction of a relationship. A similar technique, known as regression, is used to make predictions about scores on one variable from knowledge of scores on another. These predictions are obtained from the **regression line.** The regression line is the best-fitting straight line through a set of points in a scatter diagram. It is found by using the principle of least squares, which minimizes the squared deviation around the regression line.

The mean is the point of least squares for any single variable. This means that the sum of the squared deviations around the mean will be smaller than it will be at any other point. The regression line is the running mean or line of least squares in two dimensions or in the space created by two variables. Consider the situation shown in the scatter diagram in Fig. 10.1. For each level of X

(or point on the X scale) there is a distribution of scores on Y. In other words, we could find a mean of Y when X is 3 and another mean of Y when X is 4 and so on. The least squares method in regression finds the straight line that comes as close to as many of these Y means as possible. In other words, it is the line at which the squared deviations around the line are at a minimum.

The formula for a regression line is

$$Y' = a + bX \leftarrow \text{Raw score or actual value of } X \tag{10.1}$$

Predicted value of Y intercept   Regression coefficient

To use the regression equation we must define some of the terms. The term on the left of the equation is Y'. This is the predicted value of Y. When we create the equation we use observed values of Y and X. The equation is the result of the least squares procedure and shows the best available relationship between X and Y. When the equation is available we can take a score on X and plug it into the formula. What results is a predicted value of Y.

The most important term in the equation is the **regression coefficient** or b. It is the slope of the regression line. The regression coefficient can be expressed as the ratio of the sums of squares for the covariance to the sums of squares for X. It can be defined by the following formula:

$$b = \frac{N(\Sigma XY) - (\Sigma X)(\Sigma Y)}{N\Sigma X^2 - (\Sigma X)^2}. \tag{10.2}$$

The slope describes how much change is expected in Y each time X increases by one unit. For example, Fig. 10.4 shows a regression line with a slope of .67. In this figure the difference between 1 and 2 in units of X is associated with an expected difference of .67 in units of Y (for $X = 1$, $Y = 2.67$ and for $X = 2$, $Y = 3.34$; $3.34 - 2.67 = .67$). The regression coefficient is sometimes expressed in different notation. For example, the Greek β is often used for a population estimate of the regression coefficient.

The **intercept** a is the value of Y when X is 0. In other words, it is the point at which the regression line crosses the Y axis. This is shown in Fig. 10.4. It is easy to find the intercept once we know the regression coefficient. The intercept is found using the following formula:

$$a = \bar{Y} - b\bar{X}. \tag{10.3}$$

■ The Best-Fitting Line

Correlational methods require us to find the best-fitting line through a series of data points. In Fig. 10.4 a regression line is shown. This line is based on a series of observations for particular individuals. Each individual had actually obtained a score on X and on Y. Let's take the example of someone who obtained a score of 4 on X and 6 on Y. The regression equation gives a predicted

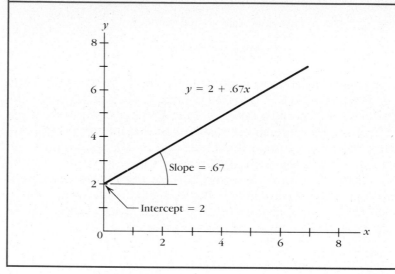

Figure 10.4

The regression equation. The slope is the amount of increase on the Y-axis divided by the range of scores on the X-axis. The intercept is the value of Y when X is 0.

value for Y that is denoted as Y'. Using the regression equation we can calculate Y' for this person. It is

$$Y' = 2 + .67X,$$

so

$$Y' = 2 + .67(4)$$
$$= 4.68.$$

If that person received a score of 6, the regression equation predicts that he would have a score of 4.68. The difference between the observed and predicted score $(Y - Y')$ is called the **residual.** The best-fitting line keeps residuals to a minimum. In other words, it attempts to find the one straight line that minimizes the deviation between observed and predicted Y-scores. Since residuals can be positive or negative, this is most appropriately done by squaring each residual. Thus, the best-fitting line is obtained by keeping these squared residuals as small as possible. This is known as the principle of least squares. Formally it is stated as

$$\Sigma(Y - Y')^2 \quad \text{is at a minimum.}$$

An example showing how to calculate a regression equation is given in Box 10.1. Whether or not you become proficient at calculating regression

• BOX 10.1

Calculation of a Regression Equation (Data from Table 10.1)

Formulas:
$$b = \frac{N(\Sigma XY) - (\Sigma X)(\Sigma Y)}{N\Sigma X^2 - (\Sigma X)^2}$$ (10.2)

$$a = \bar{Y} - b\bar{X}$$ (10.3)

STEPS

1. Find N by counting the number of pairs of observations.

   $N = 10$

2. Find ΣX by summing the X scores.

   $15 + 13 + 11 + \cdots + 2 = 78$

3. Find ΣY by summing the Y scores.

   $30 + 22 + 27 + \cdots + 16 = 213$

4. Find ΣX^2. Square each X score and then sum them.

   $225 + 169 + 121 + \cdots + 4 = 774$

5. Find ΣY^2. Square each Y score and then sum them.

   $900 + 484 + 729 + \cdots + 256 = 4729$

6. Find ΣXY. For each pair of observations multiply X by Y. Then sum the products.

   $(15 \times 30) + (13 \times 22) + (11 \times 27) + \cdots + (2 \times 16)$
   $$= 450 + 286 + 297 + \cdots + 32 = 1821$$

7. Find $(\Sigma X)^2$ by squaring the result of Step 2.

   $78^2 = 6084$

8. Find $(\Sigma Y)^2$ by squaring the result of Step 3.

   $213^2 = 45{,}369$

9. Find $N\Sigma XY$ by multiplying the results of Steps 1 and 6.

   $10 \times 1821 = 18{,}210$

10. Find $(\Sigma X)(\Sigma Y)$ by multiplying the results of Steps 2 and 3.

    $78 \times 213 = 16{,}614$

(continued)

• BOX 10.1 (*continued*)

11. Find $N\Sigma XY - (\Sigma X)(\Sigma Y)$ by subtracting the results of Step 10 from the results of Step 9.

$18{,}210 - 16{,}614 = 1{,}596$

12. Find $N\Sigma X^2$ by multiplying the results of Steps 1 and 4.

$10 \times 774 = 7{,}740$

13. Find $N\Sigma X^2 - (\Sigma X)^2$ by subtracting the results of Step 7 from those of Step 12.

$7{,}740 - 6{,}084 = 1{,}656$

14. Find b by dividing the results of Step 11 by those of Step 13.

$$\frac{1{,}596}{1{,}656} = .96$$

15. Find $\bar{X}$ by dividing Step 2 by Step 1.

$$\frac{78}{10} = 7.80$$

16. Find $\bar{Y}$ by dividing Step 3 by Step 1.

$$\frac{213}{10} = 21.30$$

17. Find $b\bar{X}$ by multiplying Step 14 by Step 15.

$.96 \times 7.8 = 7.49$

18. Find a by subtracting the results of Step 17 from Step 16.

$21.30 - 7.49 = 13.81$

19. The resultant regression equation is

$Y = a + bX = 13.81 + (.96)X.$

equations, you should learn to interpret them to be a good consumer of research information.

Correlation is a special case of regression in which the scores for both variables are in standardized or Z units. Having the scores in Z units is a nice convenience because it eliminates the need to find the intercept. In correlation the intercept will always be 0. Furthermore, the slope in correlation will be easier to interpret because it is in a standardized unit. An example of how to calculate a correlation coefficient for the data in Table 10.1 is given in Box 10.2. In the calculation of the correlation coefficient, we can bypass the step of

Calculation of a Correlation Coefficient (Data from Table 10.1)

Formula:
$$r = \frac{N\Sigma XY - (\Sigma X)(\Sigma Y)}{\sqrt{[N\Sigma X^2 - (\Sigma X)^2][N\Sigma Y^2 - (\Sigma Y)^2]}}$$

STEPS

1. Find N by counting the number of pairs of observations.

$N = 10$

2. Find ΣX by summing the X scores.

$15 + 13 + 11 + \cdots + 2 = 78$

3. Find ΣY by summing the Y scores.

$30 + 22 + 27 + \cdots + 16 = 213$

4. Find ΣX^2. Square each X score and then sum them.

$225 + 169 + 121 + \cdots + 4 = 774$

5. Find ΣY^2. Square each Y score and then sum them.

$900 + 484 + 729 + \cdots + 256 = 4729$

6. Find ΣXY. For each pair of observations multiply X by Y. Then sum the products.

$(15 \times 30) + (13 \times 22) + (11 \times 27) + \cdots + (2 \times 16)$
$$= 450 + 286 + 297 + \cdots + 32 = 1821$$

7. Find $(\Sigma X)^2$ by squaring the result of Step 2.

$78^2 = 6084$

8. Find $(\Sigma Y)^2$ by squaring the result of result of Step 3.

$213^2 = 45{,}369$

9. Find $N\Sigma XY$ by multiplying the results of Steps 1 and 6.

$10 \times 1821 = 18{,}210$

10. Find $(\Sigma X)(\Sigma Y)$ by multiplying the results of Steps 2 and 3.

$78 \times 213 = 16{,}614$

(*continued*)

11. Find $N\Sigma XY - (\Sigma X)(\Sigma Y)$ by subtracting the results of Step 10 from the results of Step 9.

$$18{,}210 - 16{,}614 = 1{,}596$$

12. Find $N\Sigma X^2$ by multiplying the results of Steps 1 and 4.

$$10 \times 774 = 7{,}740$$

13. Find $N\Sigma X^2 - (\Sigma X)^2$ by subtracting the results of Step 7 from those of Step 12.

$$7{,}740 - 6{,}084 = 1{,}656$$

14. Find $N\Sigma Y^2$ by multiplying the results of Steps 1 and 5.

$$10 \times 4{,}729 = 47{,}290$$

15. Find $N\Sigma Y^2 - (\Sigma Y)^2$ by subtracting the results of Step 8 from Step 14.

$$47{,}290 - 45{,}369 = 1{,}921$$

16. Find $\sqrt{[N\Sigma X^2 - (\Sigma X)^2][N\Sigma Y^2 - (\Sigma Y)^2]}$ by multiplying Steps 13 and 15 and then taking the square root of the product.

$$\sqrt{1{,}656 \times 1{,}921} = \sqrt{3{,}181{,}176}$$
$$= 1{,}783.59$$

17. Find

$$r = \frac{N\Sigma XY - (\Sigma X)(\Sigma Y)}{\sqrt{[N\Sigma X^2 - (\Sigma X)^2][N\Sigma Y^2 - (\Sigma Y)^2]}}$$

by dividing the results of Step 11 by Step 16.

$$\frac{1596}{1783.59} = .89$$

changing all of the scores into Z units. This gets done as part of the calculation process. You may notice that Steps 1 through 13 are identical for calculating regression (Box 10.1) and the correlation (Box 10.2). Box 10.3 gives a theoretical discussion of correlation and regression.

The **Pearson Product Moment** correlation coefficient which is explained in Box 10.2 is a ratio used to determine the degree of variation in one variable that can be estimated from knowledge about variation in the other variable. The correlation coefficient can take on any value from -1.0 and 1.0.

• BOX 10.3

A More Theoretical Discussion of Correlation and Regression

The difference between correlation and regression is analogous to that between standardized scores and raw scores. In correlation we look at the relationship between variables when each one is transformed into standardized scores. In Chapter 4 standardized or Z-scores were defined as $(X - \bar{X})/S$. In correlation, both variables are in Z-scores, so they both have a mean of 0. In other words, the mean for the two variables will always be the same. As a result of this convenience, the intercept will always be 0 (when X is 0, Y is also 0) and will drop out of the equation. The resulting equation for translating X into Y then becomes $Y = rX$. In other words, the predicted value of Y will equal X times the correlation between X and Y. If the correlation between X and Y is .80 and the standardized (Z) score for the X variable is 1.0, the predicted value of Y would be .80. Unless there is a perfect (1.0 or −1.0) correlation, scores on Y will be predicted to be closer to the Y mean than scores on X. A correlation of .80 means that the prediction for Y is 80% as far from the mean as the observation for X. A correlation of .50 would mean that the predicted distance between the mean of Y and the predicted Y is half of the distance between the associated X and the mean of X.

One of the benefits of using the correlation coefficient is that it has a reciprocal nature. The correlation between X and Y will always be the same as the correlation between Y and X. For example, if the correlation between drug dosage and activity is .68, the correlation between activity and drug dosage is .68. Regression is used to transform scores on one variable into estimated scores on the other. We often use regression to predict raw scores on Y on the basis of raw scores on X. For instance, we might seek an equation to predict grade point average on the basis of Scholastic Aptitude Test score. Because regression uses the raw units of the variables, the reciprocal prop-

erty does not hold. The coefficient describing the regression of X on Y is usually not the same as the coefficient describing the regression of Y on X.

The term *regression* was first employed in 1885 by an extraordinary British intellectual named Sir Francis Galton. Galton was fond of describing social and political changes that occur over successive generations. He noted that extraordinarily tall men tended to have sons who were a little shorter than they and that unusually small men tended to have sons closer to the average height (but still shorter than average). Over the course of time, individuals with all sorts of unusual characteristics tended to produce offspring who were closer to the average. Galton thought of this as regression toward mediocrity. This idea became the basis for a statistical procedure that describes how scores tend to regress toward the mean. If a person had been extreme on X, regression would predict that he or she would be less extreme on Y. Karl Pearson developed the first statistical models of correlation and regression in the late 19th century.

Statistical definition of regression

Regression analysis is used to show how change in one set of scores is related to change in another set of scores. In psychological testing we often use regression to determine whether changes in test scores are related to changes in performance. Do people who score higher on tests of manual dexterity perform better in dental school? Can IQ score measured during high school predict monetary income 20 years later? Regression analysis and related correlational methods tell us the degree to which these variables are linearly related. In addition, they give us an equation that estimates scores on a criterion (such as dental school grades) on the basis of scores on a predictor (such as manual dexterity).

(continued)

In Chapter 3 we introduced the concept of variance. You might remember that the variance was defined as the average squared deviation around the mean. We used the term *sum of squares* for the sum of squared deviations around the mean. Symbolically this was

$$\Sigma(X - \bar{X})^2.$$

The variance was the sum of squares divided by $N - 1$. The formula for this is

$$S_X^2 = \frac{\Sigma(X - \bar{X})^2}{N - 1}. \qquad (10.4)$$

We also gave some formulas for the variance of raw scores. The variance of X could be calculated from raw scores using the formula

$$S_X^2 = \frac{\Sigma X^2 - \frac{(\Sigma X)^2}{N}}{N - 1}. \qquad (10.5)$$

If there were another variable, Y, we could calculate the variance using a similar formula:

$$S_Y^2 = \frac{\Sigma Y^2 - \frac{(\Sigma Y)^2}{N}}{N - 1} \qquad (10.6)$$

To calculate regression we need a term for the *covariance*. This tells us how much two measures covary, or vary together. To understand covariance let's look at the extreme case of the relationship between two identical sets of scores. In this case there will be a perfect association. We know that we can create a new score which exactly repeats the scores on any one variable. If we created this new twin variable, it would covary perfectly with the original variable. Regression analysis attempts to determine how similar the variance between two variables is by dividing the covariance by the average variance from each variable.

To calculate the covariance we need to find the sum of cross products, which is defined as

$$\Sigma XY = \Sigma(X - \bar{X})(Y - \bar{Y}), \qquad (10.7)$$

and the raw score formula, which is often used for calculation, is

$$\Sigma XY - \frac{(\Sigma X)(\Sigma Y)}{N}. \qquad (10.8)$$

The covariance is the sum of cross products divided by $N - 1$.

Now look at the similarity of the formula for the covariance and the formula for the variance.

$$S_{XY}^2 = \frac{\Sigma XY - \frac{(\Sigma X)(\Sigma Y)}{N}}{N - 1} \qquad (10.8)$$

$$S_X^2 = \frac{\Sigma X^2 - \frac{(\Sigma X)^2}{N}}{N - 1} \qquad (10.5)$$

Try substituting X for Y in the formula for the covariance. You should get

$$\frac{\Sigma XX - \frac{(\Sigma X)(\Sigma X)}{N}}{N - 1}. \qquad (10.9)$$

If you replace ΣXX with ΣX^2 and $(\Sigma X)(\Sigma X)$ with $(\Sigma X)^2$ you will see the relationship between variance and covariance.

$$\frac{\Sigma X^2 - \frac{(\Sigma X)^2}{N}}{N - 1} \qquad (10.5)$$

In regression analysis we examine the ratio of the covariance to the average of the variances for the two separate measures. This gives us an estimate of how much variance in one variable we can determine by knowing about the variation in the other variable.

As you can see from Boxes 10.1 and 10.2 the calculation of the correlation coefficient and the regression equation can be a long and difficult process. You may be able to avoid the many computational steps by using a calculator. There are many inexpensive pocket calculators preprogrammed for correlation and regression. When you go to buy a calculator it would be worthwhile to hunt for one with these functions.

Testing the Statistical Significance of a Correlation Coefficient

One of the most important questions in evaluating a correlation is whether it is larger than would be expected by chance. In other words, we hope to determine whether the association between two variables is significantly different from 0. Correlation coefficients can be tested for statistical significance using the t-distribution. The formula for calculating the t-value is

$$t = r\sqrt{\frac{N - 2}{1 - r^2}}.$$ (10.10)

The significance of the t-value, where df $= N - 2$ and N is the number of pairs, can then be obtained using Appendix 2.

Let's take one example of a correlation of .37 based on 50 pairs of observations. Using the formula we obtain

$$t = .37\sqrt{\frac{48}{.86}}$$

$$= .37(7.47)$$

$$= 2.76.$$

Now, suppose we had stated the null hypothesis that the association between these two variables was not significantly different from 0. This would be tested against the alternative hypothesis that the association between the two measures was significantly different from 0 in a two-tailed test. A significance level of .05 was used. Formally, then, the hypothesis and alternative hypothesis are

H_0: $r = 0$,
H_1: $r \neq 0$.

Using the formula we obtain a t-value of 2.76 with 48 degrees of freedom. According to Appendix 2 this t-value is sufficient to reject the null hypothesis.

Thus, we conclude that the association between these two variables was not due to chance.

There are also statistical tables that give the critical values for r. One of these tables is included as Appendix 4. The table lists critical values of r for both the .05 and the .01 alpha levels according to degrees of freedom. For the correlation coefficient df $= N - 2$. Suppose, for example, that you want to determine whether a correlation coefficient of .45 is statistically significant for a sample of 20 subjects. The degrees of freedom would be 18 $(20 - 2 = 18)$. According to Appendix 4, the critical value for the .05 level is .444 with 18 df. Since .45 exceeds .444, you would conclude that the chances of finding a correlation as large as the one observed by chance alone would be less than 5 in 100. However, the observed correlation is less than the criterion value for the .01 level (that would require .561 with 18 df).

How to Interpret a Regression Plot

Regression plots are pictures that show the relationship between variables. A common use of regression is to determine the criterion validity of a test, or the relationship between a test score and some well-defined criterion. For example, the association between a test of job aptitude and the criterion of actual performance on the job is an example of criterion validity. The problems we deal with in studies of criterion validity require us to predict some criterion score on the basis of a predictor or test score. Suppose that you want to build a test to predict how enjoyable someone will turn out to be as a date. If you selected your dates randomly and had no information about them in advance, you might be best off just using normative information.

You might expect that the distribution of enjoyableness of dates is normal. Some people are just not fun for you to go out with, whereas others are exceptionally enjoyable. However, the great majority of the people are some fun but not excessively so and fall between these two extremes. Figure 10.5 shows what a frequency distribution for enjoyableness of dates might look like. As you can see in the graph, the highest point, which shows where dates are most frequently classified, is where the average date would be.

If you had no other way of predicting how much you would like your dates, the safest prediction would be to pick this middle level of enjoyableness because it is the one observed most frequently. This is called *normative* because it uses information gained from representative groups. Knowing nothing else about an individual, we can make an educated guess that a person will be average in enjoyableness because past experience has demonstrated that the mean or average score is also the one observed most frequently. In other words, knowing about the average date gives you some information about

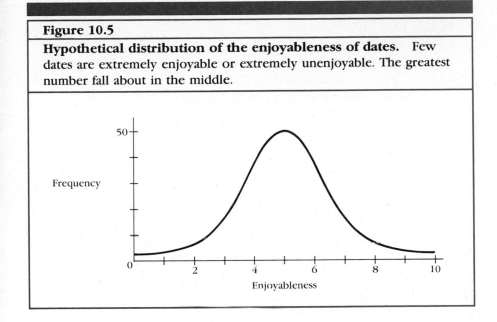

Figure 10.5

Hypothetical distribution of the enjoyableness of dates.   Few dates are extremely enjoyable or extremely unenjoyable. The greatest number fall about in the middle.

what to expect from a particular date. But it is doubtful that you would really want to choose dates this way. You probably would rather use other information such as the person's educational background, attitudes, and hobbies to predict whether that person would be enjoyable for you to spend an evening with.

Most people do employ some system to help them make important personal choices. These systems, however, are never perfect. Thus you are left with something that is not perfect but still better than just using normative information. In regression studies we develop equations that help us describe more precisely where tests fall between being perfect predictors and being no better than just using the normative information. This is done by graphing the relationship between test scores and the criterion. Then a mathematical procedure is used to find the straight line that comes as close to as many of the points as possible.

Figure 10.6 shows the points on hypothetical scales of dating desirability and the enjoyableness of dates. The line through the points is the one that minimizes the squared distance between the line and the data points. In other words, the line is the one straight line that summarizes more about the relationship between dating desirability and enjoyableness than any other straight line.

Figure 10.7 shows the hypothetical relationship between a test and a criterion. Using this figure you should be able to find the predicted value on the criterion variable by knowing the score on the test or the predictor. Here is

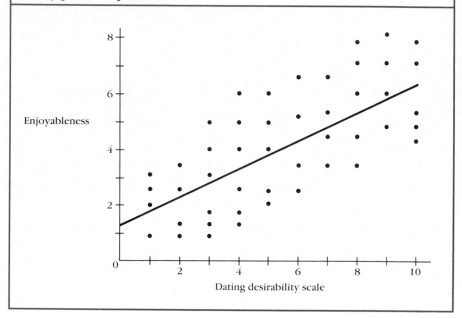

Figure 10.6

Hypothetical relationship between dating desirability and the enjoyableness of dates.   Each point summarizes the dating desirability score and enjoyableness rating for a single subject. The line was derived from a mathematical procedure to come as close to as many points as possible.

how you would read the graph. First pick a particular score on the test, say 8. Find 8 on the axis of the graph above the label "Test Score" (horizontal axis). Now draw a line straight up until you hit the slanted line on the graph. This is the regression line. Now make a 90° left turn and draw another line until it hits the other axis of the graph, which is labeled "Criterion Score." The dashed line in Fig. 10.7 shows the course you should have taken. Now read the number on the criterion axis where your line has stopped. This score of 7.4 on the criterion variable is the score you would have expected to obtain on the basis of information you gained by using the test.

Notice that the line in Fig. 10.7 is not at a 45° angle and that the two variables are measured in the same units. If it were a 45° angle, the test would be a perfect (or close to perfect) forecaster of the criterion. However, this is almost never the case in practice. Now do the same exercise you did for the test score of 8 with test scores from the extremes of the distributions. Try the scores 0 and 10. If you use these you will find that the score of 10 for the test

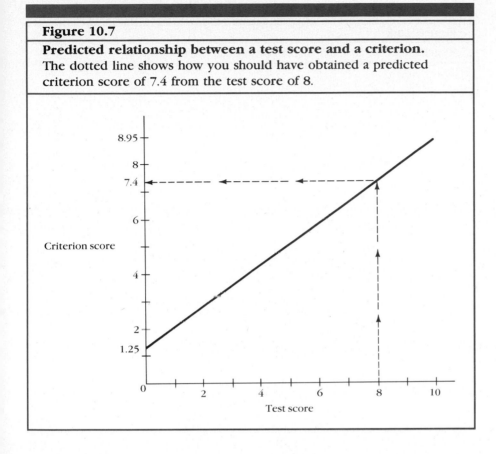

Figure 10.7

Predicted relationship between a test score and a criterion.
The dotted line shows how you should have obtained a predicted
criterion score of 7.4 from the test score of 8.

gives you a criterion score of 8.95, and the test score of 0 gives you a criterion
score of 1.25. Notice how far apart 0 and 10 are on the test. Now look at how
far apart 1.25 and 8.95 are on the criterion. You will see that using the test as a
predictor is not as good as perfect prediction, but it is still better than using
the normative information. If we had used only the normative information, we
would have predicted that all scores would be the average score on the criterion. And if there were perfect prediction, the distance between 1.25 and 8.95
on the criterion would have been the same as the distance between 0 and 10
on the test.

Figure 10.8 shows a variety of different regression slopes. Notice that the
higher the standardized regression coefficient, the steeper the line appears.
Now look at the regression line with a slope of 0. This one is parallel to the
axis for the test score and perpendicular to the line for the criterion. A regression line like this shows that the test score tells us nothing about the criterion
beyond the consensus information.

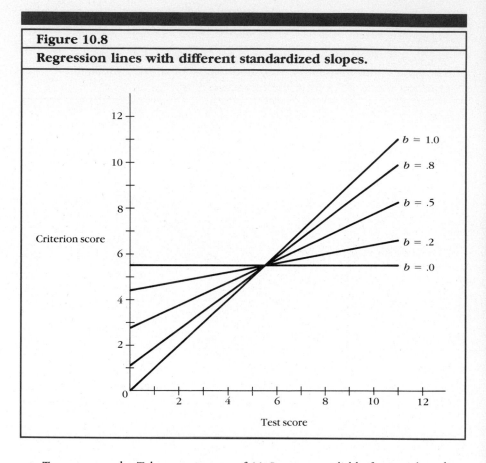

Figure 10.8

Regression lines with different standardized slopes.

Try an example. Take a test score of 11. Just as you did before, make a line straight up from this score until you hit the regression line for a slope of 0. Now draw the line at a 90° angle to the left until you hit the "Criterion Score" axis. The score should be 5.5. Now do the same thing for a test score of 3. Did you get the same predicted criterion score? You should have. In fact, any test score will give you the same predicted criterion score. And that score always will be the average score on the criterion. The slope of 0 tells you that the test and the criterion are unrelated and that your best bet under these circumstances will be to predict the average score on the criterion.

Take some time and try finding the predicted score on the criterion for test scores of 11 and 3. Do it for several of the different slopes shown in Fig. 10.8. Notice that the steeper the slope of the regression line, the farther apart the predicted scores on the criterion. Table 10.2 shows the predicted scores for all the different slopes. You can use it to check your answers.

When the regression lines have slopes of 0 or nearly 0, it is best not to take any chances in forecasting the criterion. Instead you should depend on the

Table 10.2		
Expected criterion scores for two test scores when predicted from regression lines with different slopes		
TEST SCORE	SLOPE	PREDICTED CRITERION SCORE
11	1.0	11.00
3	1.0	3.00
11	.8	9.90
3	.8	3.50
11	.5	8.25
3	.5	4.25
11	.2	6.60
3	.2	5.00
11	.00	5.50
3	.00	5.50

normative information and guess the mean of Y. As the slope becomes steeper, we are more prepared to take some chances and estimate that there will be differences in criterion scores.

It is instructive to think about Fig. 10.8 as we reflect on the meaning of psychological tests. For example, if the SAT has a slope of .5 for predicting grades in college, this would mean that the SAT performance relationship is defined by the line labeled "$b = .5$" in Fig. 10.8. Using this sort of information college administrators can forecast that there will be some slight differences in college performance which might be predicted from the SAT. However, because the slope is not steep, those predictions are somewhere between perfect prediction and what they would get if they used the consensus information.

Other Correlation Coefficients

The Pearson Product Moment correlation is only one of many types of correlation coefficient. It is the one most commonly used because most often we want to find the correlation between two continuous variables. Continuous variables can take on any values over a range of values. Height, weight, and intelligence are examples of continuous variables. However, there are situations where we want to find the correlations between variables which are scaled in other ways.

Spearman's rho is a method of correlation that is used to find the association between two sets of ranks. The rho coefficient (ρ) is easy to calculate and is often used when the individuals in a sample can be ranked on two variables but their actual scores are not known.

There is a whole family of correlation coefficients that involves dichotomous variables. Dichotomous variables have only two levels. Examples might be yes-no or male-female. Some dichotomous variables are called "true dichotomous" because they naturally form two categories. For example, sex is a true dichotomous variable and has the two levels male and female. Other dichotomous variables are called "artificial dichotomous" because there was originally an underlying continuous scale which was forced into a dichotomy. Passing or failing a bar examination might be an example of an artificial dichotomy. (Many scores can be obtained, but the examiners decide only to consider pass or fail.) The types of correlation coefficients used to find the relationship between dichotomous and continuous variables are shown in Table 10.3.

The *biserial correlation* expresses the relationship between a continuous variable and an artificial dichotomous variable. For example, the biserial correlation might be used to assess the relationship between passing or failing the bar examination (artificial dichotomous variable) and grade point average in law school (continuous variable). If the dichotomous variable had been "true" (such as sex), we would use the *point biserial correlation*. For instance, the point biserial correlation would be used to find the relationship between sex and grade point average. When both variables are dichotomous and at least one of the dichotomies is "true," the association between them can be estimated using the *phi coefficient*. For example, the relationship between passing or failing the bar examination and sex (male or female) could be estimated using the phi coefficient. If both dichotomous variables are artificial, we might

Table 10.3			
Appropriate correlation coefficients for relationships between dichotomous and continuous variables*			
VARIABLE *Y*	VARIABLE *X*		
	Continuous	Artificial dichotomous	True dichotomous
Continuous	Pearson *r*	Biserial *r*	Point biserial *r*
Artificial dichotomous	Biserial *r*	Tetrachoric *r*	Phi
True dichotomous	Point biserial *r*	Phi	Phi

*The entries in the table suggest which type of correlation coefficient is appropriate given the characteristics of the two variables. For example, if variable *Y* is continuous and variable *X* is true dichotomous, you would use the point biserial correlation.

employ a special correlation coefficient known as the *tetrachoric correlation*. Among these special correlation coefficients, the point biserial, phi, and Spearman's rho coefficients are probably used most often. The formulas for calculating these types of correlation are given in Box 10.4.

● BOX 10.4

Formulas for Spearman's Rho, the Point Biserial Correlation, and the Phi Coefficient

Spearman's Rho:
$$\rho = 1 - \frac{6\Sigma d_i^2}{N^3 - N} \qquad (10.11)$$

where

ρ = Spearman's rho coefficient
d_i = a subject's rank order on variable 2 minus his/her rank order on variable 1, and
N = number of paired ranks.

When used: To find the association between pairs of observations, each expressed in ranks.

Point Biserial Correlation:
$$r_{\text{pbis}} = \left[\frac{\bar{Y}_1 - \bar{Y}}{S_y} \right] \sqrt{\frac{P_x}{(1 - P_x)}} \qquad (10.12)$$

where

r_{pbis} = the point biserial correlation coefficient,
X = a true dichotomous (two-choice) variable,
Y = a continuous (multilevel) variable,
$\bar{Y}_1$ = the mean of Y for subjects having a "plus" score on X,
$\bar{Y}$ = the mean of Y for all subjects,
S_y = the standard deviation for Y scores, and
P_x = the proportion of subjects giving a "plus" score on X.

When used: To find the association between a dichotomous (two-choice) variable and a continuous variable. For the true dichotomous variable, one of the two choices is arbitrarily designated as a "plus" response.

Phi Coefficient:
$$\phi = \frac{P_c - P_x P_y}{\sqrt{P_x(1 - P_x)P_y(1 - P_y)}} \qquad (10.13)$$

(*continued*)

• BOX 10.4 (*continued*)

where

ϕ = the phi coefficient,
P_c = the percentage in the "plus" category for both variables,
P_x = the percentage in the "plus" category for the first variable, and
P_y = the percentage in the "plus" category for the second variable.

When used: To find the association between two dichotomous (two category) variables. A dichotomous variable might be "yes-no" or "on-off." In each case one of the two choices is arbitrarily chosen as a "plus" response. When using phi, one of the variables must be a "true" dichotomy (if both were "artificial" the tetrachoric correlation would be more appropriate).

Terms and Issues in the Use of Correlation

When you use correlation or read studies that report correlational analysis, you will need to know the terminology. Some of the terms and issues you should be familiar with include *residual, standard error of estimate, coefficient of determination, coefficient of alienation, shrinkage, cross validation, correlation causation problem,* and *third variable.* Following are brief discussions of these terms and concepts.

RESIDUAL.   A regression equation gives a predicted value of Y' for each value of X. In addition to these predicted values, there are observed values of Y. The difference between the predicted and the observed values is called the *residual.* Symbolically, the residual is defined as $Y - Y'$.

In regression analysis the residuals have certain properties. One important property is that the sum of the residuals will always equal 0, $[\Sigma(Y - Y') = 0]$. In addition, the sum of the squared residuals will be the smallest value according to the principle of least squares $[\Sigma(Y - Y')^2 = \text{smallest value}]$.

STANDARD ERROR OF ESTIMATE.   Once we have obtained the residuals we can find their standard deviation. However, in creating the regression equation two constants (a and b) have been found. Thus we must use two degrees of freedom rather than one, as is usually the case in finding the standard deviation. The standard deviation of the residuals is known as the *standard error of estimate,* which is defined as

$$S_{yx} = \sqrt{\frac{\Sigma(Y - Y')^2}{N - 2}}.$$ (10.13)

The standard error of estimate is a measure of the accuracy of prediction. Prediction is most accurate when the standard error of estimate is relatively small. As it becomes larger, the prediction becomes less accurate.

COEFFICIENT OF DETERMINATION. The square of the correlation coefficient is known as the *coefficient of determination.* This value tells us the percentage of total variation in scores on Y that we know as a function of information about X. For example, if the correlation between the SAT and performance in the first year of college is .40, the coefficient of determination would be .16. This means that we could explain 16% of the variation in first-year college performance as a function of knowing SAT scores.

COEFFICIENT OF ALIENATION. The *coefficient of alienation* is a measure of nonassociation between two variables. This is calculated as $\sqrt{1 - r^2}$, where r^2 is the coefficient of determination. For the SAT example, the coefficient of alienation would be $\sqrt{1 - .16} = \sqrt{.84} = .92$. This means that there is a high degree of nonassociation between the SAT and college performance.

SHRINKAGE. Many times a regression equation is created on one group of subjects and then used to predict the performance of another group. One of the problems with regression analysis is that it takes advantage of chance relationships within a particular population of subjects. Thus, there is a tendency to overestimate the relationship, particularly if the sample of subjects is small. *Shrinkage* is the amount of decrease observed when a regression equation is created for one population and then applied to another. Formulas are available to estimate the amount of shrinkage to expect given characteristics of the variance, covariance, and sample size (Lord, 1950; McNemar, 1969; Uhl and Eisenberg, 1970).

An example of shrinkage might be when a regression equation is developed to predict first-year college grade point average on the basis of the Scholastic Aptitude Test. Although the percentage of variance in grade point average might be fairly high for the original group, we can expect to account for a smaller percentage of the variance when the equation is used to predict grade point average in the next year's class. This decrease in the percentage of variance accounted for is the shrinkage.

CROSS VALIDATION. The best way to ensure that proper inferences are being made is to use the regression equation to predict performance in a

• BOX 10.5

The Danger of Inferring Causation from Correlation

A recent article published in a newspaper discussed the stressfulness of a variety of occupations. A total of 130 job categories were rated for stressfulness by examining Tennessee hospital and death records for evidence of stress-related diseases such as heart attacks, ulcers, arthritis, and mental disorders. The 12 highest and the 12 lowest were as follows.

MOST STRESSFUL	LEAST STRESSFUL
1. Unskilled laborer	1. Clothing sewer
2. Secretary	2. Garment checker
3. Assembly-line inspector	3. Stock clerk
4. Clinical lab technician	4. Skilled craftsperson
5. Office manager	5. Housekeeper
6. Foreperson	6. Farm laborer
7. Manager/administrator	7. Heavy-equipment operator
8. Waitress/waiter	8. Freight handler
9. Factory machine operator	9. Childcare worker
10. Farm owner	10. Factory package wrapper
11. Miner	11. College professor
12. House painter	12. Personnel worker

The article advises readers who want to remain healthy to avoid the "most stressful" job categories.

group of subjects other than the ones on which the equation was created. Then a standard error of estimate can be obtained for the relationship between the values predicted by the equation and those observed. This process is known as *cross validation.*

THE CORRELATION-CAUSATION PROBLEM.   Just because two variables are correlated, it does not necessarily mean that one has caused the other (see Box 10.5). For example, because there is a correlation between the number of hours spent viewing television and aggressive behavior, it does not mean that excessive viewing of television causes aggression. This relationship could mean that an aggressive child might prefer to watch a lot of television.

THIRD VARIABLE EXPLANATION.   In the example of television and aggression there are other possible explanations for the observed relationship between viewing and aggressive behavior. One is that some third variable causes both excessive viewing of television and aggressive behavior. For example, poor social adjustment might explain both. Thus the apparent relationship

However, the evidence may not warrant the authors' advice. Although it is quite possible that diseases are associated with particular occupations, this does not necessarily mean that holding the jobs causes the illnesses. There are a variety of other explanations. For example, people with the propensity for heart attacks and ulcers might be more likely to select jobs as unskilled laborers or secretaries. Thus the direction of causation might be that health condition causes job selection rather than the reverse. Another possibility is a third variable explanation. Some other factor might cause the apparent relationship between job and health. For example, the income level might cause both stress and illness. It is well known that poor people have lower health status than wealthy people. It is possible that impoverished conditions might cause a person to accept certain jobs and also to have more diseases.

These three possible explanations are now diagrammed. An arrow shows a causal connection.

		Economic Status
		↙ ↘
Job → Illness	Job ← Illness	Job Illness
Job causes illness	Tendency toward illness causes people to choose certain jobs	Economic status (third variable) causes job selection and illness

In this example we are *not* ruling out the possibility that job causes health condition. In fact, it is quite plausible. However, because the nature of the evidence is correlational, it cannot be said with certainty that job causes illness.

between viewing and aggression actually might be the result of some variable that is not included in the analysis and may be unknown. We usually refer to this external influence as a *third variable*.

Summary

This chapter began with a discussion of a claim made in the *National Enquirer* that poor diet causes marital problems. Actually there was no specific evidence that diet *causes* the problems—only that diet and marital difficulties are associated. However, the *Enquirer* failed to specify the exact strength of the association. The rest of the chapter was designed to help you be more specific than the *Enquirer* article was by learning to specify associations with precise mathematical indexes known as *correlation coefficients*.

First we presented pictures of the association between two variables, which are called *scatter diagrams*. Second we presented a method for finding a linear equation to describe the relationship between two variables. This regression method uses the data in raw units. The results of regression analysis are two constants: a *slope* describing the degree of relatedness between the variables and an *intercept* giving the value of the *Y* variable when the *X* variable is 0. When both of the variables are in standardized or *Z* units, the intercept is always 0, and it drops out of the equation. In this unique situation we solve for only one constant, which is *r,* or the *correlation coefficient.*

When using correlational methods we must take many things into consideration. For example, correlation does not mean the same thing as causation. In the case of the *Enquirer* article the observed correlation between diet and problems in marriage may mean that diet causes the personal difficulties. However, it may also mean that marriage problems cause poor eating habits or that some *third variable* causes both diet habits and marital problems. In addition to the difficulties associated with causation, we must always consider the strength of the correlational relationship. The *coefficient of determination* describes the percentage of variation in one variable which is known on the basis of its association with another variable. The *coefficient of alienation* is an index of what is not known from information about the other variable.

Key Terms

1. **Scatter Diagram**   A picture of the relationship between two variables. Individual data points are plotted on a two dimensional graph.
2. **Correlation Coefficient**   A mathematical index that describes the direction and magnitude of a two-dimensional relationship.
3. **Positive Correlation**   A statistical association in which higher scores on one variable are associated with higher scores on the other variable.
4. **Negative Correlation**   A statistical association in which higher scores on one variable are associated with lower scores on the other variable.
5. **Regression Line**   The best-fitting straight line through a set of points in a scatter diagram. The regression line is described by a mathematical index known as the regression equation.
6. **Regression Coefficient**   The ratio of covariance to variance. Also known as the slope of the regression line. The regression coefficient describes how much change is expected in the *Y* variable each time the *X* variable increases by one unit.
7. **Intercept**   The value of *Y* when *X* is zero.
8. **Residual**   The difference between the predicted value given by a regression equation and the observed value.
9. **Standard Error of Estimate**   The standard deviation of the residuals obtained from the regression equation.

10. **Coefficient of Determination** The correlation coefficient squared. This is an estimate of the percentage of total variation of scores on Y that can be explained as a function of variation on X.

11. **Coefficient of Alienation** A measure of nonassociation between two variables. It is given by $\sqrt{1 - r^2}$, where r^2 is the coefficient of determination.

Exercises

The following table gives scores for reading and mathematics for 16 third grade classes in the San Diego Unified School District.

School	Reading	Mathematics
Adams	240	257
Bayview	269	271
Carson	225	220
Crown Point	237	326
Edison	221	221
Fletcher	274	276
Fremont	329	309
Kennedy	202	220
La Jolla	260	280
Lowell	251	294
Marvin	329	336
Oak Park	288	293
Penn	262	257
Sherman	200	224
Toler	391	385
Whittier	251	260

1. Calculate the regression equation for predicting reading from mathematics scores.

2. Would you typically use regression for this sort of problem?

3. Calculate the correlation between reading and mathematics.

4. Is the relationship between reading and mathematics positive or negative?

5. How much of the variation in reading scores is accounted for by knowing mathematics scores?

6. What is the name of the index used to estimate this percentage?

7. Does mathematics ability cause students to read well, or is it their reading ability that causes them to do well in mathematics?

11

Reliability and Validity

Correlation and regression are among the most widely used statistical methods. It is not uncommon for research investigators to cite correlation coefficients when talking about their work. They usually expect that you know what they mean when they say that there is a moderate correlation of .34 between two variables.

It would not be possible to discuss all of the applications of correlation and regression in a single chapter. There are literally thousands of them. Instead, this chapter focuses on applications of correlational analysis in a particular area—**psychometrics**. Psychometrics is the study of mental measurements. Experts in psychometrics, known as psychometricians, spend their careers developing and using psychological tests. Many of the principles of psychometrics are applicable to other areas of inquiry. Psychometricians have developed sophisticated methods to quantify characteristics of measurements, and the importance of these contributions is becoming recognized in many other academic disciplines. Research in education has applied these principles for many years, and they are now finding greater use in medicine, public health, physical education, sociology, anthropology, biology, and economics. This list is not exhaustive. Indeed, psychometric methods are becoming basic scientific tools.

This chapter reviews some of the basic methods for assessing reliability and validity.* Typically, these topics are not taught in introductory statistics courses, but they are a very important aspect of most behavioral and health science research. Many serious problems with current research stem from inadequate understanding of some of the basic principles of statistics. The review in

*Several examples in this chapter are taken from Kaplan and Saccuzzo (1982) and are reproduced by permission of the authors and Brooks/Cole Publishing Co.

this chapter is not complete since it is only intended to provide some basic background.

Measurement error is common in all fields of science. Psychological, health, and educational specialists have devoted a great deal of time and study to measurement error and its effects. Tests that are relatively free of measurement error are considered to be reliable, and tests that contain relatively great measurement error are considered to be unreliable. The basics of contemporary theories and methods of reliability were worked out in the early part of this century by Charles Spearman. Test score and reliability theories have gone through continual refinements since then.

When we evaluate reliability we must first specify the source of the measurement error that we are trying to evaluate. There are many possible sources of error. For example, error can result from the test being given at different times, from selecting a small sample of items to represent a larger conceptual domain, or creating a test from items that are unrelated to one another. There are different methods for each of these sources of error and each is presented in this chapter.

Discrepancies between true and measured scores constitute errors of measurement. In statistics, we do not attach a negative connotation to the word *error*. It does not imply that a mistake has been made. Rather we acknowledge that there will always be some inaccuracy or error in our measurements. Our task is to find the magnitude of this error and to develop ways to minimize it. This chapter is about the conceptualization and assessment of measurement error.

Basics of Test Score Theory

The classical test score theory assumes that each person has a true score that would be obtained if there were no errors in measurement. However, because our measuring instruments are imperfect, the score we observe for each person may differ from the person's true ability or characteristic. The difference between the true score and the score we observe results from **measurement error**. In symbolic representations, we can say that the observed score X is composed of two components: a true score T and an error component E:

$$X = T + E,$$ (11.1)

observed score — true score — error

or we can say that the difference between the score we obtain and the score we are really interested in is equal to the error of measurement:

$$E = X - T.$$ (11.2)

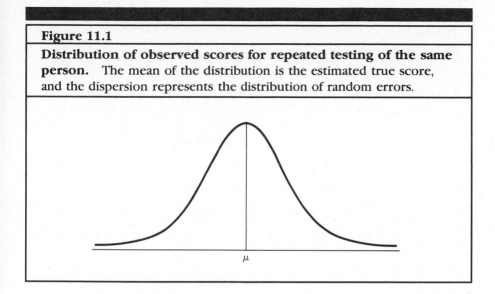

Figure 11.1

Distribution of observed scores for repeated testing of the same person. The mean of the distribution is the estimated true score, and the dispersion represents the distribution of random errors.

μ

A major assumption in classical test theory is that errors of measurement are random. Although systematic errors are acknowledged in most measurement problems, they are less likely to force an investigator to make the wrong conclusions than are random errors. A carpenter who always misreads a tape measure by 2 inches (or makes a systematic error of 2 inches) would still be able to cut boards to be the same length. Working with an unreliable measure might be like a carpenter using a rubber yardstick. If the test makes systematic errors it might be analogous to a carpenter who works with a ruler that is always stretched to be 2 inches too long. Classical test theory deals with the rubber yardstick problems in which the ruler stretches and contracts in a random way.

Using a random rubber yardstick we would not get the same score on each measurement. Instead we would get a distribution of scores like that shown in Fig. 11.1. Each observed score in the distribution would be composed of two components: a true score component T and an error component E. Basic sampling theory tells us that the distribution of random errors would be bell-shaped. Thus the center of the distribution should represent the true score, and the dispersion around the mean of the distribution should display the distribution of sampling errors. This tells us that any one application of the rubber yardstick may not tell us the true score, but through repeated applications we would be able to estimate the true score by finding the mean of the observations.

Figure 11.2 shows three different distributions. In the first distribution there is a great dispersion around the true score. In this case you might not want to depend on a single observation because it might be quite far from the

Figure 11.2

Three distributions of observed scores.   The left distribution is one in which error is the greatest; the right distribution has the least error.

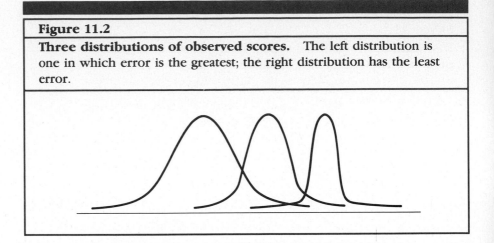

true score. The third distribution in Fig. 11.2 displays a situation in which the dispersion around the true score is very small. In this case most of the observations are very close to the true score, and it might be accurate to draw conclusions on the basis of fewer observations.

The dispersions or distributions around the true score in Figs. 11.1 and 11.2 tell us how much error there is in the measure. Classical test theory assumes that the true score for an individual will not change with repeated applications of the same test. Because of random error, however, repeated applications of the same test can produce different scores. Theoretically, the standard deviation of the distribution of errors tells us about the magnitude of measurement error. Although this theory only deals with the distribution of errors for a single person, we usually assume that the distribution of random errors will be the same for all persons. Thus, classical test theory uses the standard deviation of errors as the basic measure of error. Usually, we can call this the **standard error of measurement** and describe it symbolically as

σ_{meas}.

The rubber yardstick analogy may be useful in helping you understand this concept. Suppose that you have a table that is 30 inches high. If you were to measure the height of the table using a steel yardstick, you would probably always find it to be the same height—30 inches. However, now you try to measure the table with the rubber yardstick. The first time you try, the stick has stretched and you record 32 inches. The next time you discover that the stick has shrunk, and you record 28 inches for the height of the table. Now we are in trouble because repeated applications of the yardstick will not always give us the same information about the height of the table.

However, there is one way out of this situation. If we assume that the yardstick stretches and shrinks in a random fashion, then we can say that the distribution of scores given by the yardstick will be normal. Most scores will be close to the actual or true score, and we would expect scores that are greatly different from the true score to occur less frequently. Thus, it would be rare to observe a score as low as 5 inches or as high as 55 inches. The mean of the distribution of scores from repeated applications of the rubber yardstick would be an estimate of the true height of the table. The standard deviation would be the standard error of measurement. Remember from Chapter 3 that the standard deviation tells us something about the average deviation around the mean. The standard error of measurement tells us, on the average, how much a score varies from the true score.

There are several methods for estimating reliability and each of these requires the use of the correlation coefficient. Typically reliability is covered in courses about psychological testing. Only a few methods of reliability assessment are covered here.

Sources of Error

There may be many reasons why an observed score is different from a true score. There may be situational factors such as loud noises in the room while a test is being administered. The room may be too hot, or it may be too cold. The health status of the test-takers also could affect test scores. For example, you may know how hard it is to do well on an examination when you have a cold or are just feeling depressed. Reliability assessment deals with these factors directly by using time sampling. Here the test is given at different points in time. Each test administration is considered an independent sample. The most common use of time sampling is **test-retest reliability**. With this method the same group of people is tested with the same test at two different times. Then the correlation between these two samples is used as an estimate of reliability. Time-related error is only one of many sources of error. Reliability theory is also used to evaluate errors that are internal or related to the correlations between the items of the test. Internal problems in a test are studied using internal consistency methods. Methods for assessing each of these types of errors follow.

Test-Retest Method

Test-retest reliability estimates are used to evaluate the error associated with administering a test at two different times. This type of analysis is only of value if we are measuring traits or characteristics of individuals that are not believed to change over time. We usually assume that an intelligence test measures a general ability that is not transient. Therefore, if we administer an IQ test at two different times and get different scores, we might conclude that the lack of correspondence is due to random measurement error. Usually, we do not assume that a person got smarter or less smart.

Test-retest reliability is relatively easy to evaluate. All you need to do is administer the same test on two well-specified occasions and then find the correlation between the scores using the correlational methods presented in Chapter 10. The resulting correlation is considered an index of reliability.

General Measures of Internal Consistency

There are many methods for estimating the internal consistency of a test. One simple method is to divide the test into halves and to find the correlation between these two subtests. This is known as the split-half method. Many years ago, Kuder and Richardson (1937) greatly advanced reliability assessment by developing methods for evaluating reliability within a single administration of a test, but their approach does not depend on some arbitrary splitting of the test into halves.

Decisions about how to split tests into halves cause many potential problems for split-half reliability. The two halves of the test may have different variances. For example, the split-half method requires that each half of the test be scored separately, and this can create additional work. The Kuder-Richardson technique avoids these problems because it is a general method that simultaneously considers all the possible ways of splitting the items.

■ The KR$_{20}$ Formula

The formula for calculating the reliability of a test in which the items are scored 0 or 1 (usually for right or wrong) is known as the KR$_{20}$. The formula came to be labeled this way because it was the 20th formula presented in a famous article by Kuder and Richardson (the K and R are obtained from the authors' initials). The formula is

$$\text{KR}_{20} = \frac{N}{N-1}\left(\frac{S^2 - \Sigma\,pq}{S^2}\right), \tag{11.3}$$

where

S^2 = the variance of the total test scores,
p = the proportion of people getting each item correct,
q = the proportion of people getting each item incorrect,
N = the number of items on the test, and
$\Sigma\,pq$ = the sum of the product of p and q for each item on the test.

Although we will not go through the mathematical derivations of the KR_{20}, studying the components of the formula may give you a better understanding of how it works. (An example of the calculation is given in Box 11.1.) First you will recognize the term S^2 from Chapter 3. This is the variance of the test scores, and it appears twice in the formula. The term Σpq is the sum of the proportion of people passing each item times the proportion of people failing each item. It can be shown that the product pq is the variance for an individual item. Thus, Σpq is the sum of the individual item variances.

• BOX 11.1

The Calculation of Reliability Using KR_{20}

Formulas:
$$KR_{20} = \frac{N}{N-1}\left(\frac{S^2 - \Sigma pq}{S^2}\right)$$

$$S^2 = \frac{\Sigma X^2 - \dfrac{(\Sigma X)^2}{NS}}{NS - 1}$$

where: NS = number of test-takers = 50*
         N = number of items = 6

Data:

ITEM	NUMBER OF TEST-TAKERS RESPONDING CORRECTLY	p (FROM STEP 2)	q (FROM STEP 3)	pq (FROM STEP 4)
1	12	.24	.76	.18
2	41	.82	.18	.15
3	18	.36	.64	.23
4	29	.58	.42	.24
5	30	.60	.40	.24
6	47	.94	.06	.06
				$\Sigma pq = \overline{1.10}$ (from Step 5)

STEPS

1. Determine the number of test-takers NS.

   $NS = 50$

2. Find p by dividing the number of people responding correctly to each item by the number of people taking the test (Step 1). This is the level of difficulty.

   $\dfrac{12}{50} = .24$ $\dfrac{41}{50} = .82$ $\cdots$

3. Find q for each item by subtracting p (the result of Step 2) from 1.0. This gives the proportion responding incorrectly to each item.

   $1.0 - .24 = .76 \qquad 1.0 - .82 = .18 \qquad \cdots$

4. Find pq for each item by multiplying the results of Steps 2 and 3.

   $(.24)(.76) = .18 \qquad (.82)(.18) = .15 \qquad \cdots$

5. Find Σpq by summing the results of Step 4 over the N items.

   $.18 + .15 + .23 + .24 + .24 + .06 = 1.1$

6. Find S^2, which is the variance for the test scores. To do this you will need the scores for each individual in the group. The formula for the variance is

$$S^2 = \frac{\Sigma X^2 - \left[\dfrac{(\Sigma X)^2}{NS}\right]}{NS - 1}.$$

   In this example $S^2 = 2.8$. (This is given for this example.)

7. Find $S^2 - \Sigma pq$ by subtracting the results of Step 5 from Step 6.

   $2.8 - 1.1 = 1.7$

8. Find $(S^2 - \Sigma pq)/S^2$ by dividing the results of Step 7 by Step 6.

   $\dfrac{1.7}{2.8} = .607$

9. Find N, the number of items.

   $N = 6$

10. Find $N/(N - 1)$ by dividing the results of Step 9 by Step 9 minus 1.

   $\dfrac{6}{5} = 1.2$

11. Find KR_{20} by multiplying the results of Steps 8 and 10.

   $(1.2)(.607) = .73$

---

*Although number of subjects is often denoted by N, we use NS in this example, where we are using N to denote number of items.

Now let us think about conditions that would make the term on the right side of the equation either large or small. First consider the situation in which the variance is equal to the sum of the variances of the individual items. Symbolically this would be

$$S^2 = \Sigma pq.$$

In this case, the right side of the KR_{20} formula would be 0, and as a result the estimate of reliability would be 0. This tells us that to have nonzero reliability, the variance for the total test score must be greater than the sum of the variances for the individual items. The only situation that will make the total test score variance more than the sum of the item variance is when there is *covariance* between the items. Covariance occurs when the items are correlated with one another. The greater the covariance, the smaller the Σpq term will be. When the items covary they can be assumed to measure the same general trait, and the reliability for the test will be high. As Σpq approaches 0, the right side of the equation approaches 1.0. The other factor in the formula is an adjustment for the number of items in the test, which allows the estimate of reliability to be higher for larger tests. An example of the calculation of KR_{20} is given in Box 11.1.

■ Coefficient Alpha

Mathematical proofs have demonstrated that the KR_{20} formula gives the same estimate of reliability as taking the mean of the split-half reliability estimates obtained by dividing the test in all possible ways (Cronbach, 1951). You can see that the Kuder-Richardson procedure is very general and usually will be more valuable than a split-half estimate of internal consistency. However, there are still cases in which the KR_{20} formula will not be appropriate for evaluation of internal consistency. These situations occur when a test has items that are not scored as "right" or "wrong."

There are many types of tests for which there are no right or wrong answers. This is typically the case for personality and attitude scales. For example, on an attitude questionnaire you might be presented with a statement such as "I believe extramarital sex is immoral." You must indicate whether you: strongly disagree, disagree, are neutral, agree, or strongly agree. None of these choices is incorrect, and none of these is correct. Rather, your response indicates where you stand on the continuum between agreement and disagreement. To extend the Kuder-Richardson method for use with this sort of item, Cronbach developed a formula that could be used to estimate the internal consistency of tests where the items are not scored as 0 or 1 (right or wrong). In doing so, Cronbach developed a more general reliability estimate, which he called coefficient alpha or α. The formula for coefficient alpha is

$$\alpha = \left(\frac{N}{N-1} \right) \left(\frac{S^2 - \Sigma S_i^2}{S^2} \right). \tag{11.4}$$

As you may notice, this looks very similar to the KR_{20} formula. The only difference is that Σpq has been replaced by ΣS_i^2. This new term is for the variance of the individual items. The summation sign informs us that we are to sum the individual item variances. The only real difference is the way the variance of the items is expressed. Coefficient alpha is a more general reliability coefficient than KR_{20} because S_i^2 can describe the variance of items whether or not they are in a right-wrong format. Thus coefficient alpha is the most general case of internal consistency reliability.

All of the measures of internal consistency evaluate the extent to which the different items on a test measure the same ability or trait. They all will give low estimates of reliability if the test is designed to measure several traits.

One of the factors that affects the internal consistency estimate of a test is the number of items. If all items are measures of the same trait or ability, simply increasing the test length will improve the reliability. But how many items should be added? The prophecy formula can be used to make an estimate.

■ The Prophecy Formula

The **prophecy formula** is used for estimating how long a test must be to achieve a desired level of reliability. The formula is

$$N = \frac{r_d(1 - r_o)}{r_o(1 - r_d)},$$
(11.5)

where

N = the number of tests that are the length of the current version needed to bring it up to the desired level of reliability,

r_d = the desired level of reliability, and

r_o = the observed level of reliability based on the current version of the test.

Consider an example in which a 20-item test had a reliability of .76. We would like to bring the reliability up to .85. Putting these numbers into the prophecy formula, we get

$$N = \frac{.85(1 - .76)}{.76(1 - .85)} = \frac{.204}{.114} = 1.79.$$

These calculations tell us that we would need 1.79 tests the length of the current 20-item test to bring the reliability up to the desired level. To find the number of items that would be required for the test, we must multiply the number of items on the current test by N from the formula. In the example, this would give

$$20 \times 1.79 = 35.8.$$

So the test would have to be expanded from 20 to about 36 items to achieve the desired reliability of .85. This assumes that the added items come from the same pool as the original items and that they have the same psychometric properties.

The decision to expand a test from 20 to 36 items depends on economic and practical considerations. The test developer first must ask whether the increase in reliability is worth the extra time, effort, and expense required to achieve this goal. If the test is to be used for personnel decisions it may be dangerous to ignore any effort that will enhance the reliability of the test. On the other hand, if the test is only to be used to get an idea of whether two variables are associated, the expense of extending it may not be worth the effort and cost.

When the prophecy formula is used, certain assumptions are made that may or may not be valid. One of these assumptions is that the probability of error in items added to the test is the same as that for the original items in the test. However, there may be some occasions in which adding many items would bring about new sources of error. For instance, if the test becomes very long, test-taker fatigue might become a major source of error.

As an example of a situation in which it might not be worthwhile to attempt to bring up the reliability of a test, consider a 40-item test that has a reliability of .50. We would like to bring the reliability up to .90. Using the prophecy formula, we get

$$N = \frac{.90(1 - .50)}{.50(1 - .90)} = \frac{.90(.50)}{.50(.10)} = \frac{.45}{.05} = 9.$$

These figures tell us that the test would have to be 9 times its present length to have a projected reliability of .90. This means it must have

9 × 40 = 360 items.

Creating a test 360 items long would be very expensive, and validating it would require a considerable amount of time for both test-makers and test-takers. Beyond these problems, there might be new sources of error in the 360-item test that would not have been a problem for the shorter one. For example, many errors may occur on the longer test simply because people get tired and bored during the long process of answering 360 questions. There is no way of taking these factors into account using the prophecy formula.

Validity

Validity can be defined as the agreement between a test score or measure and the quality it is believed to measure. Over the years psychologists have created many subcategories of validity. The most commonly discussed are: *content, criterion,* and *construct.* Content validity describes the extent to which test items represent a conceptual domain. This is largely a logical process that does not require statistical analysis. Correlational analysis is most often used to assess criterion and construct validity.

■ Validity Coefficients in Criterion Validity

The relationship between a test and a criterion is usually expressed as a correlation. When used in this context, the correlation is called a **validity coefficient,** and it tells the extent to which the test is valid for making statements about the criterion.

Not all validity coefficients have the same value, and there are no hard-and-fast rules about how large the coefficient must be to be meaningful. In practice it is rare to see a validity coefficient larger than .60, and validity coefficients in the range of .30 to .40 are commonly considered acceptable. A coefficient is statistically significant if the chances of obtaining its value by chance alone are quite small (usually less than 5 in 100). For example, suppose that the Scholastic Aptitude Test (SAT) had a validity coefficient of .40 for predicting grade point average at a particular West Coast university. Because this coefficient is likely to be statistically significant, we can say that the SAT score tells us more about how well people will do in college than we would know by chance.

There are many reasons why college students differ from one another in their academic performance. You probably could easily list a dozen. Because there are so many factors that contribute to college performance, it would be too much to expect that the SAT could explain all of the variation. The question we must ask is, "How much of the variation in college performance will we be able to predict on the basis of SAT scores?"

The validity coefficient squared is the percentage of variation in the criterion that we can expect to know in advance because of our knowledge of the test scores. Thus we will know $.40^2$, or 16%, of the variation in college performance because of the information we have from the SAT test. The remainder of the variation in college performance (84%) is still unexplained. In other words, when students arrive at college, most of the reasons why they perform differently will still be a mystery to college administrators and professors.

However, if school officials have used the SAT to select the students, they did so to learn about 16% of the variation between students that could be predicted before any students are admitted. It is important to realize that accounting for small percentages of the variance in some criterion under different circumstances is very useful for people making decisions. Obviously, college administrators feel that the small bit of information they get from SAT scores aids them in selecting the right students for their institution.

There are many circumstances in which using a test may not be worth the effort if it only contributes a few percentage points to the understanding of variation in a criterion. Actually moderate validity coefficients (.30 to .40) sometimes can be very useful even though they may explain only about 10% of the variation in the criterion. For example, Dunnette (1976) demonstrated how a simple questionnaire for military screening could save taxpayers $4 million every month even though the validity was not remarkably high. (This analysis was done before the days of high inflation. If it were done today, the estimated savings would be much greater!) There also are some circumstances in which a validity coefficient of .30 or .40 means almost nothing.

■ Construct Validity

Before 1950 most social scientists considered only criterion and content forms of validity. By the mid-1950s investigators concluded that no clear criteria existed for most of the social and psychological characteristics they wanted to measure. Developing a measure of intelligence, for example, was difficult because no one could say for certain what intelligence was. Criterion validity studies would require that a specific criterion of intelligence be established against which the tests could be compared. The problem was that there was no criterion for intelligence. Intelligence is actually a hypothetical construct we cannot touch or feel.

In contemporary psychology we often want to measure intelligence, love, curiosity, or mental health. These are constructs that are not clearly defined, and there are no established criteria against which we can compare the accuracy of our tests. These are the truly challenging problems in measurement.

Construct validity is a series of activities in which a researcher simultaneously defines some construct and develops the instrumentation to measure it. This process is required when "no criterion or universe of content is accepted as entirely adequate to define the quality to be measured" (Cronbach and Meehl, 1955). Construct validation involves assembling evidence about what a test really means. This is done by showing the relationship between a test and other tests and measures. Each time a relationship is demonstrated, one additional bit of meaning can be attached to the test. Over a series of studies, the meaning of the test gradually begins to take shape. Construct validation is an ongoing process, similar to amassing support for a complex scientific theory. No single set of observations provides crucial or critical evidence. Yet over time many observations gradually clarify what the test means.

Several years ago Campbell and Fiske (1959) introduced an important set of logical considerations for developing tests with construct validity. They distinguished between two types of evidence that are essential for a meaningful test. These forms of evidence are convergent and discriminant. To argue that a test has meaning, a test constructor is advised to be armed with as much evidence as possible for these two types of validity.

■ Convergent Evidence

Convergent evidence for validity is obtained when a measure correlates well with other tests that are believed to measure the same construct. This sort of evidence shows that measures of the same construct "converge" or narrow in on the same thing. In many ways convergent evidence for construct validity is like criterion validity; in each case scores on the test are related to scores on some other measure. Criterion validity is fine for situations in which you are attempting to predict performance on a particular variable—such as success in graduate school. Here the task is well-defined, and all you need to do is find the items that are good predictors of this graduate school criterion. In construct validity there is no well-defined criterion, and the meaning of the test comes to be defined by the variables it can be shown to be associated with.

An example of the need for construct validation evidence comes from studies by Kaplan, Bush, and Berry (1976), who attempted to define and measure the construct "health" by developing a health index. Because of the complexity of the health concept, no single measure can serve as the criterion against which a measure of health can be assessed. This situation requires studies in construct validation. Some of the construct validation studies were used to demonstrate the convergent validity of the health index.

Convergent evidence is obtained in one of two ways. The first is by showing that a test measures the same things as other tests used for the same purpose. The second is by demonstrating specific relationships that could be expected if the test is really doing its job. Studies on the health index included both types of evidence. In demonstrating the meaning of the health index the authors continually asked themselves, "If we were really measuring health, what relationships would we expect to observe between the health index and other measures?" The simplest relationship is between health index scores and the way people rate their own health status. This relationship was strong and clearly showed that the index captured some of the same information individuals used to evaluate their own health. However, a good measure must go beyond this simple bit of validity evidence because self-ratings are unreliable. If they were not, we would use self-perceived health status as the index of health because it is easier than using the health index.

However, in construct validity no single variable can serve as the criterion. Thus other studies were used to show a variety of other relationships. For example, it was shown that people who scored as less healthy on the health index also tended to report more symptoms and chronic medical conditions. It also was hypothesized that health status would be related to age, and it was observed that these two variables were indeed systematically related; older persons tended to be lower in health status.

The researchers also evaluated specific hypotheses that were advanced on the basis of some theoretical notion about the construct. In the health index studies the authors reasoned, "If the index really measures health, then we would expect that people who score low on the measure should visit doctors more often." A study confirmed that they do indeed, and this provided evidence for one more inference. Also, certain groups (such as disabled persons) should have lower average scores on the index than other groups (such as nondisabled persons). Again, a study confirmed this hypothesis (Kaplan, Bush, and Berry, 1976). Thus in a series of studies the number of meanings that could be given to the health index was gradually expanded. Yet convergent validity does not comprise all of the evidence that is necessary to argue for the meaning or construct validity of a psychological test or measure. We also must have discriminant evidence for construct validity.

■ Discriminant Evidence

Science can be very conservative. It confronts scientists with difficult questions such as, "Why should we believe your theory if we already have a theory that seems to say about the same thing?" An eager scientist may answer this question by arguing that his or her theory is distinctive and better. In testing we have a similar dilemma. Why should we create a new test if there is already one available to do the job? Thus one type of evidence a person needs in test validation is proof that the test measures something unique. For example, if a health index measures the same thing as self-rated health, number of symptoms, and number of chronic medical conditions, why do we need it in addition to all these other measures? The answer is that the measure taps something other than the tests used in the convergent validity studies. This demonstration of uniqueness is called **discriminant evidence**.

By providing evidence that a test measures something different from other tests we also are measuring some unique construct. Calling an old construct by a new name always should be avoided. Discriminant evidence indicates that the measure does not represent a construct other than the one for which it is devised.

The Relationship Between Reliability and Validity

In validity studies, the meaning of a test comes to be defined by demonstrating correlations between the test and other variables. Attempting to define the validity of a test will be a futile effort if the test is not reliable. Theoretically, a test should not correlate more highly with any other variable than it correlates with itself. The maximum validity coefficient between two variables is equal to the square root of the product of their reliabilities. Or:

$$r_{12} \max = \sqrt{r_{11} r_{22}},\tag{11.6}$$

where

$r_{12} \max$ = the maximum correlation, and
r_{11} and r_{22} = the reliabilities of the first and second tests.

For example, the maximum correlation between two tests with reliabilities of .8 and .5, respectively, is

$$r_{12} \text{ max} = \sqrt{(.8)(.5)}$$
$$= \sqrt{.40}$$
$$= .63.$$

Because validity coefficients are usually not exceptionally high, it is possible that a modest correlation between the true scores on two traits would be missed if the test for each of the traits was not highly reliable. Table 11.1 shows the maximum validity you would expect to find given various levels of reliability for two tests. Sometimes it is not possible to demonstrate that a reliable test has meaning. In other words, it is possible to have reliability without validity. However, it is logically impossible to demonstrate that an unreliable test is valid.

Table 11.1

How reliability affects validity*

RELIABILITY OF TEST	RELIABILITY OF CRITERION	MAXIMUM VALIDITY (CORRELATION)
1.0	1.0	1.00
.8	1.0	.89
.6	1.0	.77
.4	1.0	.63
.2	1.0	.45
.0	1.0	.00
1.0	.5	.71
.8	.5	.63
.6	.5	.55
.4	.5	.45
.2	.5	.32
.0	.5	.00
1.0	.0	.00
.8	.0	.00
.6	.0	.00
.4	.0	.00
.2	.0	.00
.0	.0	.00

*The first column shows the reliability of the test. The second column displays the reliability of the validity criterion. The numbers in the third column are the maximum theoretical correlations between tests, given the reliability of the measures.
(*Source:* Kaplan and Saccuzzo, 1982.)

Correction for Attenuation

Low reliability is a real problem in psychological research because it reduces the chances of finding significant correlations between measures. We know that if a test is unreliable, information obtained with it can be of little or no value. Thus we say that potential correlations are *attenuated* or diminished by measurement error.

Fortunately, measurement theory does allow us to estimate what the correlation between two measures would have been if they had not been measured with error. These methods "correct" for the attenuation in the correlations caused by the measurement error. To use the methods one needs to know only the reliabilities of two tests and the correlation between them. We estimate $\hat{r}_{12}$ (or "r-hat" for the notation above the r), which is the estimated true correlation. The **correction for attenuation** is

$$\hat{r}_{12} = \frac{r_{12}}{\sqrt{r_{11}r_{22}}},$$ (11.7)

where

$\hat{r}_{12}$ = the estimated true correlation between Tests 1 and 2,
r_{12} = the observed correlation between Tests 1 and 2,
r_{11} = the reliability of Test 1, and
r_{22} = the reliability of Test 2.

Suppose, for example, that a test of manual dexterity and a test of general athletic skill were correlated .34, and the reliabilities of the tests were .75 and .82, respectively. The estimated true correlation between athletic ability and manual dexterity is

$$\frac{.34}{\sqrt{(.75)(.82)}} = \frac{.34}{\sqrt{.615}} = \frac{.34}{.78} = .44.$$

As the example shows, the estimated correlation increases from .34 to .44 when the correction is used.

Sometimes one measure meets an acceptable standard of reliability but the other one does not. In this case we would want to correct for the attenuation caused only by the unreliable test. To do this we use the formula

$$\hat{r}_{12} = \frac{r_{12}}{\sqrt{r_{11}}},$$ (11.8)

where

$\hat{r}_{12}$ = the estimated true correlation,
r_{12} = the observed correlation, and
r_{11} = the reliability of the variable that does not meet our standard of reliability.

For example, suppose we want to estimate the correlation between manual dexterity and grade point average in dental school. The reliability of the manual dexterity test is .75, which is not acceptable to us, but dental school GPA is assumed to be measured without error. Using the fallible manual dexterity test, we observed the correlation to be .53. Plugging these numbers into the formula, we get

$$\frac{.53}{\sqrt{.75}} = \frac{.53}{.87} = .61.$$

This informs us that the correcting for the attenuation caused by the manual dexterity test increased our observed correlation from .53 to .61.

A word of caution about correction for attenuation is necessary. The lower the level of reliability, the less meaningful these corrections will be. The methods are most useful for estimating how high a correlation could have been "potentially." They do not substitute for real observations.

Summary

Measurement error is common to all forms of science. Methods for evaluating measurement error and its consequences have been developed by psychometricians, who are mental-measurement experts. These methods are now finding widespread use in many areas of science. The two most important psychometric properties of a score or measure are reliability and validity.

Theoretically, reliability is the ratio of true variance to observed variance. In practice, reliability is estimated using correlation coefficients. Two general classes of reliability are considered in this chapter: test-retest and internal consistency. Test-retest reliability is estimated from the correlation between scores on a particular test that was administered at two different points in time. One method for evaluating the internal consistency of a test requires that the test be split in half and that a correlation between the two halves be computed. This method, known as the split-half reliability, is not optimal because is depends on the way that the test has been arbitrarily split. More sophisticated methods, such as the KR_{20} formula, give the average correlation between all possible divisions of the test items. However, the KR_{20} method can only be used for tests with items scored right or wrong. A more general method, known as coefficient alpha, can be used for scaled responses as well as for right-wrong items.

Validity defines the meaning of a test or a measure. Two types of validity that use correlation coefficients are criterion validity (based on the correlation between a test and some well-defined criterion) and construct validity. Construct validity is studied when the domain of measurement is not well defined. In the case of construct validity, development of the measure and definition of the construct occur simultaneously.

Reliability and validity are related. A test can be reliable without being valid. However, a test cannot be valid without being reliable. The reason for this is that low reliability reduces the correlation between the test and other measures. If a test has very low reliability, it will not correlate with any other test or measure. Since validity is demonstrated by correlations, it is impossible to show meaningful associations with the unreliable test.

There are some methods for estimating the impact of low reliability on correlation coefficients. Correction for attenuation can be used to correct for unreliability and to estimate what the correlation would have been if variables had been reliable.

Key Terms

1. **Psychometrics**   The study of mental measurement.
2. **Measurement Error**   The discrepancy between a true score and a measured score.
3. **Standard Error of Measurement**   The standard deviation of the errors of measurement in a psychological test.
4. **Test-Retest Reliability**   The correlation between scores on the same test administered at two points in time.
5. **Split-Half Reliability**   An estimate of the reliability of a test based on a division of the test items into halves. A split-half reliability is the correlation between these two halves of the test.
6. **KR$_{20}$**   A formula that estimates the internal consistency of a test. This method estimates the average correlation across all different divisions of the test into two scores.
7. **Coefficient Alpha**   A general measure of internal consistency that can be used with items in any format.
8. **Prophecy Formula**   A formula that estimates how long a test must be to achieve a desired level of reliability.
9. **Validity**   The agreement between a test score or measure and the quality it is believed to measure.
10. **Validity Coefficient**   The correlation between a test score and some criterion or external measure it is purported to represent.
11. **Construct Validity**   A series of activities in which a researcher simultaneously defines some construct and develops the instrumentation to measure it. This is composed of two types of evidence: *convergent evidence* showing that the test measures something similar to other measures of the same construct and *discriminant evidence* indicating that the test is distinct and measures something different than other tests.
12. **Correction for Attenuation**   A mathematical correction that adjusts the correlation when one or both variables is measured with error.

Exercises

1. A new IQ test was given to 12 fifth graders during the first and sixth week of school. Scores for the children are presented in the following table. Find the test-retest reliability for the new test.

Initials	First Test	Second Test
DK	105	107
DW	99	106
AW	112	105
SH	116	115
TC	111	114
CD	86	87
LS	104	96
JD	102	95
HG	87	92
HS	120	114
GT	100	104
MC	106	110

2. In your own words, describe the concept of measurement error.

3. A test with 10 items was given to 100 students. The number of students getting each item correct is presented below. Use the KR_{20} formula to calculate the internal consistency of the test. The variance of the test was 3.2.

Item	Number of Test-Takers Responding Correctly	Item	Number of Test-Takers Responding Correctly
1.	41	6.	71
2.	86	7.	16
3.	16	8.	25
4.	42	9.	91
5.	98	10.	46

4. A test with 40 items has a reliability of .81. You would like to bring the reliability of the test up to .90. How many items will you have to add?

5. The correlation between two tests was .63. However, one test had a reliability of .70 while the other had a reliability of .59. If you were to correct for the measurement error in these two tests, what would be the correlation between the two tests?

6. Suppose a researcher administered a social support test with a reliability of .5 and wanted to correlate it with blood pressure known to have a reliability of about .6. Even if the two measures were truly perfectly correlated, what level of correlation would you expect to observe?

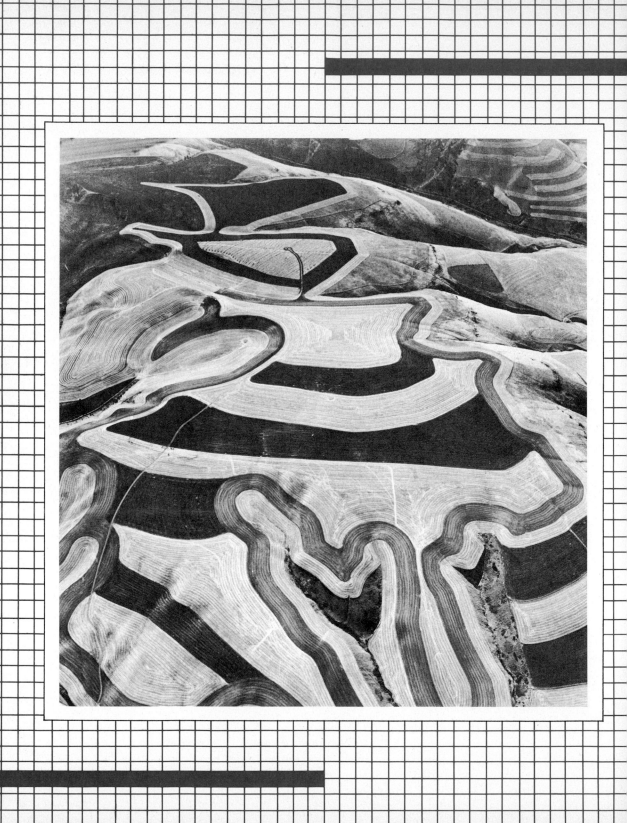

12

Chi-Square and Other Nonparametric Statistics

The preceding chapters have provided you with a wide range of statistical techniques. They cover methods of scaling, computation of the mean, the standard deviation, and the standard error of the mean. These statistics are used to describe samples of data. The mean and the standard deviation provide a simple summary of a complex set of observations. Other chapters describe methods for comparing samples of data. These methods include correlational techniques, t-tests, and analysis of variance. These are the core methods that are used most frequently in behavioral and health sciences research.

Despite the many advantages of the methods covered so far, they are not applicable in all situations. This chapter provides an introduction to two other classes of methods that you will need to interpret the research literature. First, chi-square methods are discussed, and then some examples of other nonparametric statistics are presented. Chi-square methods are used to evaluate frequency data. Parametric statistical tests such as the t-test and the analysis of variance make assumptions about the population distribution of the data. Nonparametric statistics are used when we cannot be certain that these assumptions are met.

Chi-Square for
Frequency Data

In Chapter 2 four different scales of measurement were presented: nominal, ordinal, interval, and ratio. If this seems unfamiliar to you, please go back and review pages 17–20 to refresh your memory. Chapter 2 also discussed frequency distributions. The remainder of the book has been devoted to methods for describing data that are at least ordinal. However, we frequently have data that are at the nominal level.

Nominal data are unscaled. In other words, the numbers used in nominal scales are not meaningful. They might be considered analogous to the numbers on the backs of baseball players. The numbers simply label the players, and taking an average of them would not be meaningful. A common example of a nominal scale used in social science research is for religious affiliation. Survey respondents indicate their religious preferences, and these preferences are coded according to some scheme. For example, 1 may be used for Protestant, 2 for Catholic, 3 for Jew, 4 for Moslem, 5 for Hindu, and so on. As a statistical consultant, I have frequently had people bring computer printouts showing that the mean for religion is 1.7 (with a standard deviation of .68). What does this mean? The answer is it doesn't mean anything. Although the computer will calculate a mean and a standard deviation for religion, the numbers were chosen arbitrarily and do not form any type of scale.

This does not imply that collecting information about religion is meaningless. In fact, information about ethnic and religious background is very important in our understanding of many social, behavioral, and health phenomena. What is of interest is the frequency distribution for these categories, in other words, the proportions of people who occupy each category. For example, we may wish to know whether or not the distribution of religious preferences in a small town that always votes Republican is the same as the distribution of religious preferences in the nation as a whole. Frequency data also can be used to answer questions about contingency. Sometimes we may wish to learn about the frequency of developing heart disease in each religious category. It has been suggested that consumption of red meat may be related to the development of heart disease. Since Seventh Day Adventists do not eat red meat, we may wish to know whether the chances of developing heart disease are significantly lower for Seventh Day Adventists than they are for those in other religious categories.

The **chi-square test** is used to evaluate this type of problem. A variety of different questions can be evaluated using the chi-square test. Here are some examples:

• Are more men than women opposed to nuclear war?

- Did a football team win a significant number of games beyond what would be expected by chance?
- When given an alternative, are psychology students more likely to take an elective in sociology or political science?

The Chi-Square Statistic

The chi-square statistic is used to evaluate the relative frequency or the proportions of events in a population that fall into well-defined categories. For each category, there is an expected frequency that is obtained from knowledge of the population or from some other theoretical perspective. There is also an observed frequency for each category. The observed frequency is obtained from observations made by the investigator. The chi-square statistic expresses the discrepancy between the observed and the expected frequencies. The formula for chi-square is

$$\chi^2 = \sum \frac{(O - E)^2}{E},\qquad\qquad(12.1)$$

where O is the observed frequency and E is the expected frequency.

Before using the chi-square statistic, several conditions and limitations must be considered. These are summarized in Table 12.1. The first condition is that the data must be raw frequencies. Sometimes, the data may be proportions. If this is the case, multiply the proportions for each category by the number in the sample to obtain raw category frequencies. The second condition is that the frequencies must be independent. For example, if we are comparing the number of men and women opposed to nuclear war, each person can express only one opinion. In other words, a woman cannot be counted as both favoring and not favoring this attitude. A third assumption is that the sum of the expected frequencies must be equal to the sum of the observed frequencies. And the fourth assumption is that the expected frequencies for each condition should be greater than 5. The chi-square distribution is not accurate for cases with very small expected frequencies (that is, less than 5). However, for chi-square problems with a large number of categories (4 or more) small expected frequencies do not pose a severe problem.

An example of the calculation of chi-square is presented in Box 12.1. This example compares the frequency of psychology majors choosing an elective course in either political science or sociology. This example was taken from a West Coast college that has a very strict schedule for its psychology majors. During their sophomore year, psychology majors must take either Introduction to Sociology or Introduction to Political Science. The college discourages students from selecting both courses and allows credit for graduation for only one of the two. Therefore, virtually no students take both courses, and we will

Table 12.1

Conditions and limitations for the use of chi-square

CONDITION	EXPLANATION
1. Frequency data.	Data must be raw frequencies. Chi-square values are influenced by the magnitude of scores, and proportions should not be used. If your data are proportions, multiply each proportion by the number of cases to get raw frequencies.
2. Independence.	Each event must be independent. For example, a person can be counted in one and only one category.
3. Sum of the expected frequencies must equal sum of the observed frequencies.	Always check to be certain that the sums of the expected and observed frequencies are equal.
4. Expected frequencies within each category (or cell) should be larger than 5.	Chi-square tables are not accurate when the expected frequencies are very small. This problem is less severe when there are many different categories or cells. However, if you have 4 or fewer cells, you may reach an inaccurate conclusion if one or more of your expected frequencies is less than 5.

assume for the purposes of this example that each student selects one or the other. The committee that designed the curriculum assumed that half the students would select sociology and the other half would select political science. However, they had never determined whether this assumption was correct.

The null hypothesis can be stated as

$$H_0: \quad P_1 = P_2,$$

where P_1 is the proportion of students choosing sociology and P_2 is the proportion choosing political science. This null hypothesis is tested against the alternative hypothesis

$$H_1: \quad P_1 \neq P_2.$$

A chi-square test was used to evaluate this assumption. Chi-square was chosen because

• the data are in frequency form;

Calculation of Chi-Square

H_0: $P_1 = P_2$ (where P_1 is the proportion selecting sociology and P_2 is the proportion selecting political science)

H_1: $P_1 \neq P_2$

Summary table

TYPE	SOCIOLOGY	POLITICAL SCIENCE	TOTAL
O	251	199	450
E	225	225	450
$O - E$	26	-26	0
$(O - E)^2$	676	676	1352
$\dfrac{(O - E)^2}{E}$	3.00	3.00	

Formula:
$$\chi^2 = \sum \left[\frac{(O - E)^2}{E} \right] = 3.00 + 3.00 = 6.00$$
$$df = 1$$

STEPS

1. Write down the observed frequencies for each category.

   Sociology: 251
   Political Science: 199

2. Obtain the expected frequencies for each category. In this example, there were 450 students, and we assumed that half of the students would select Sociology while the other half would select Political Science.

   Sociology: .5(450) = 225
   Political Science: .5(450) = 225

3. Obtain the total for both the expected and the observed. Check to be sure that they are equal.

   Total Observed = 251 + 199 = 450
   Total Expected = 225 + 225 = 450
   450 = 450

4. Find $O - E$ for the first category (Sociology).

   251 − 225 = 26

5. Find $O - E$ for the second category (Political Science).

   199 − 225 = −26

(continued)

● BOX 12.1 (*continued*)

6. Square the results of Step 4 to obtain $(O - E)^2$ for the first category.

$$26^2 = 676$$

7. Square the result of Step 5 to determine $(O - E)^2$ for the second category.

$$26^2 = 676$$

8. Obtain $(O - E)^2/E$ for the first category by dividing Step 6 by the expected (in this case 225).

$$\frac{676}{225} = 3.00$$

9. Find $(O - E)^2/E$ for the second category.

$$\frac{676}{225} = 3.00$$

10. Now χ^2 can be calculated by summing Steps 8 and 9.

$$\chi^2 = \sum \left[\frac{(O - E)^2}{E} \right]$$
$$= 3.00 + 3.00$$
$$= 6.00$$

11. Determine the degrees of freedom.

$$df = \text{categories} - 1$$
$$= 2 - 1$$
$$= 1$$

12. Use Appendix 5 to determine whether the observed chi-square value exceeds the criterion values for statistical significance.

For df = 1 at the .05 level, χ^2 must exceed 3.84.

$$6.00 > 3.84.$$

Conclusion: Reject null hypothesis. Significantly more than half of the psychology majors select sociology.

- the event was independent for each student (each student selected one course or the other but not both); and
- the expected frequency for both categories was greater than 5.

An overview of the calculation of chi-square is presented in the summary table in Box 12.1. First the observed frequencies for each category must be determined. At this college, there were 450 psychology majors observed over the course of three years. Among these students, 251 selected sociology and 199 selected political science. The expected frequencies for each category were obtained from the curriculum committee's assumption that half of the students would select sociology while the other half would select political science. Since there were 450 students, the committee expected 225 to select each of these courses.

The chi-square test assumes that the sum of the observed frequencies will equal the sum of the expected frequencies. This assumption must be checked. The sum of the observed frequencies is 251 + 199 = 450; the sum of the expected frequencies is 225 + 225 = 450. Therefore, the sum of the observed frequencies is equal to the sum of the expected frequencies.

The next step is to obtain the observed minus the expected frequency for each category. For sociology this is 251 − 225 = 26. For political science it is

$199 - 225 = -26$. Now the $(O - E)^2$ values are obtained. These are 676 for both the sociology and the political science category. Notice that the values for each category are equal. This will always be the case when chi-square is used to compare the frequency of two categories and when the expected value for these two categories is equal to 50% of the study sample.

The next step is to obtain $(O - E)^2/E$ for each category. As the example shows, for each category this is $676/225 = 3.00$. Chi-square χ^2 is the sum of this quantity across categories. So $\chi^2 = 3.00 + 3.00 = 6.00$. The degrees of freedom for single-variable chi-square are obtained by determining the number of categories and subtracting one. In this example, there are 2 categories, so $df = 2 - 1 = 1$. Appendix 5 gives the critical values, or chi-square, at the 5% and 1% significance levels. In this example, there was 1 degree of freedom. Appendix 5 shows that the critical value for the 5% significance level is 3.84 while the critical value for the 1% significance level is 6.64. The observed chi-square value of 6.00 exceeds the critical value for the 5% significance level. It was expected that an equal number of students would select the sociology and the political science course. However, a higher proportion of students actually selected the sociology course and the probability is less than 5 in 100 that this difference in frequency would have occurred by chance alone. Therefore, we reject the null hypothesis in favor of the alternative hypothesis.

■ A Four-Category Example

The method described in the preceding section can be easily extended to a larger number of categories. In this example we consider four categories. The data were taken from a hypothetical survey of football fans during the first week of 1984. At that point, four NFL professional football teams remained in the running for the Superbowl. The teams were the Seahawks, Raiders, 49ers, and Redskins. A summary of the analysis is presented in Table 12.2. The calculations are not broken down step by step as they were in Box 12.1. See if you can work out the calculations for Table 12.2 without using a step-by-step breakdown.

The null hypothesis is

$$H_0: \quad P_1 = P_2 = P_3 = P_4,$$

where

P_1 is the proportion favoring team 1 (Seahawks),
P_2 is the proportion favoring team 2 (Raiders),
P_3 is the proportion favoring team 3 (49ers), and
P_4 is the proportion favoring team 4 (Redskins).

This is evaluated against the alternative hypothesis

$$H_1: \quad H_0 \text{ is false.}$$

Table 12.2

Distribution of Superbowl preferences during the first week of 1984

	SEAHAWKS TEAM 1	RAIDERS TEAM 2	49ers TEAM 3	REDSKINS TEAM 4	TOTAL
1. Observed (O)	18	30	19	33	100
2. Expected (E)	25	25	25	25	100
3. $O - E$	−7	5	−6	8	0
4. $(O - E)^2$	49	25	36	64	194
5. $\dfrac{(O - E)^2}{E}$	1.96	1	1.44	2.56	6.96

$$\chi^2 = \sum \left[\frac{(O - E)^2}{E} \right]$$

$$\chi^2 = 6.96$$

Critical value at .05 level = 7.81

The first row of Table 12.2 lists the observed frequencies for football fans selecting each of the four teams as their Superbowl favorites. Eighteen people thought the Seahawks would win the Superbowl; 30 thought the Raiders would win; 19 preferred the 49ers; and 33 selected the Redskins. The second row of Table 12.2 lists the expected frequencies. In this example we say that each team has an equal expected frequency of winning the championship. Therefore, we have divided the expected frequencies among the four teams. To obtain these expected frequencies, we divide 100 by 4 and obtain 25 for each team. Row three in the table gives the observed minus the expected frequencies. Look, for example, at the expected frequency for the Seahawks. This is the observed frequency (18) minus the expected frequency (25), which is −7. Now look at the same value for the Raiders. Here the observed frequency was 30, and the expected frequency was 25. So, $O - E = 5$. The observed minus expected frequencies for the 49ers and the Redskins were −6 and 8, respectively.

At this point it is valuable to check the "Total" column in Table 12.2. The first thing to look for is whether the observed and the expected frequencies are equal. Note that the total of the observed frequencies is 100 and the total of the expected frequencies is 100. Thus, the problem meets condition three described in Table 12.1 that the observed and expected frequencies must be equal. Now look at the value in the "Total" column for the third row ($O - E$). This value is 0. In fact, the total of the observed minus the expected frequencies for all chi-square problems of this nature should be 0. If it is not, the sum of the observed frequencies is probably not equal to the sum of the expected frequencies. Another potential problem is that you have made an error in subtracting the expected from the observed frequencies. The total value is not particularly important for row four of the table. However, the total value for row five is extremely important. This is the chi-square value. Remember that chi-square is

$$\chi^2 = \sum \frac{(O - E)^2}{E},$$ (12.1)

so chi-square for this problem is equal to

$1.96 + 1 + 1.44 + 2.56 = 6.96.$

The next step would be to find the number of degrees of freedom. Recall that degrees of freedom are obtained by subtracting 1 from the number of categories, or

df = number of categories − 1. (12.2)

So in this case $df = 4 - 1 = 3$. To determine whether the observed chi-square is statistically significant, it is necessary to consult Appendix 5. Let's say that we had decided on the 5% level of significance prior to performing this analysis. Check Appendix 5 to determine whether our observed result is statistically significant. This can be done by locating the row in the table associated with 3 degrees of freedom. Now move across the row until you reach the column labeled 5%. The value there is 7.81. This means that a chi-square value must exceed 7.81 to be statistically significant. The obtained value, 6.96, is less than 7.81. Therefore, we retain the null hypothesis and conclude that the observed differences in frequencies are not statistically significant. In other words, we conclude that the observed differences in frequencies can be attributed to chance. The differences in frequencies for selecting the four teams as most likely to win the Superbowl are not significantly different. Notice that there are actually differences among the number of fans preferring each team. However, the conservative rules of science suggest that we attribute these differences to chance unless we have clear and convincing evidence that the observed differences are not the result of sampling error. In this example, the evidence is not strong enough to make a convincing (scientific) argument that the differences are not due to chance.

Chi-Square for the 2 × 2 Contingency Table

One of the most common uses of the chi-square statistic is to test for differences among groups. In an earlier chapter you learned that the t-test is one statistical method that is often used to compare two groups. In other chapters analysis of variance methods were used to compare different groups. However,

the *t*-test and the analysis of variance are used when the dependent variable is scaled. You may also want to know if two groups differ for the frequency with which some event occurs. This chapter limits comparisons to those between two groups and two response categories. This case, with two groups and two categories, can be described by a 2 × 2 contingency table. In more advanced courses you will learn methods for evaluating contingency tables with more groups and more categories. These methods are extensions of those developed for the simple 2 × 2 case.

The question evaluated in the 2 × 2 chi-square test is whether the two variables are independent of one another. In this example we are asking, "Is support of the treaty for freezing the testing and deployment of nuclear weapons independent of the respondent's sex?" If equal proportions of men and women favor the treaty, we will observe independence. However, if the proportions of men and women favoring the treaty differ significantly, we will not observe independence.

The null hypothesis is

H_0: The two variables are independent or not associated with one another.

The alternative hypothesis is

H_1: The two variables are dependent, or associated.

There are two different methods for calculating the chi-square from the 2 × 2 contingency table. The first method is illustrated in Box 12.2. The data for this example were adapted from a national poll that asked whether people favor a treaty freezing the testing and deployment of nuclear weapons. A total of 514 interviews was conducted with 246 men and 268 women. We are interested in determining whether men and women differ in their attitude toward the treaty. The data reported in the newspaper are printed in the first portion of Box 12.2. Seventy-two percent of the male respondents and 82% of the female respondents favored the treaty. For simplicity, all other responses to the item were grouped together. Thus, 28% of the men and 18% of the women gave a response other than favoring the treaty. In summary, each man and each woman either favored the treaty or gave another response. More women than men favored the treaty. The question is whether this difference between men and women is statistically significant.

To analyze this problem using the chi-square test, it is necessary to transform data into raw frequencies. Remember that Table 12.1 stated that the chi-square test requires raw frequencies. The first section of Box 12.2 gives only percentages. Each unique combination of sex and response category is called a *cell*. There are four cells in the table: Men Favor, Women Favor, Men Other, Women Other. To obtain the raw frequency for each cell, multiply the percentage for that cell by the number of people in that group. For example, to obtain a raw frequency for men favoring the treaty, multiply .72 (the proportion of men favoring the treaty) by 246 (the number of men). This gives approximately 177. (Technically .72 × 246 = 177.12. However, it is appropriate to round to the nearest whole person.) The raw frequencies for the other cells are summarized in the next portion of Box 12.2.

The next step is to obtain the expected frequencies for each cell in the table. This will require a review of some of the theory behind the chi-square test. Chi-square is a test of independence. In the 2 × 2 case we are asking whether two variables are independent of one another. In this particular example, the research question is whether the frequency of favoring the nuclear treaty is the same for men as it is for women. If the two variables are independent, the frequency of favoring the treaty does not depend on the sex of the respondent. On the other hand, if the two variables are not independent then the proportion favoring the treaty will differ between the two sexes.

The expected frequencies for each of the four cells must be derived theoretically. According to our null hypothesis the expected frequencies for men and women should be the same. In other words, the null hypothesis assumes that men and women do not differ. Therefore, using an estimate of frequency for men and women combined will explain both men and women. This may be best explained through example. First, let us determine the proportion of

● BOX 12.2

Chi-Square for 2 × 2 Contingency Table

Question: Do you favor a treaty freezing testing and deployment of nuclear weapons?

H_0:   Favoring the treaty is independent of sex.
H_1:   Favoring the treaty is not independent of sex.

Raw Data: 514 interviews
246 with men
268 with women

Percentage Data:

	FAVOR TREATY	OTHER RESPONSE	TOTAL
MEN	72%	28%	100%
WOMEN	82%	18%	100%

Transformation into Raw Frequencies:

	FAVOR TREATY	OTHER RESPONSE	TOTAL
MEN	.72 × 246 = 177	.28 × 246 = 69	246
WOMEN	.82 × 268 = 220	.18 × 268 = 48	268
TOTAL	397	117	514

Obtain Expected Frequencies:

STEPS

1.   Get column fractions. The column fractions are the proportion of the total responses that occur in each column category. The column fraction for "Favor Treaty" is

$$\frac{397}{514} = .77$$

For other responses, it is

$$\frac{117}{514} = .23$$

2.   Obtain expected cell frequencies by adjusting for column fractions. First, do this for the "Favor Treaty" column. For men the expected frequency would be the column fraction times the number of men, or

.77 × 246 = 189;

for women,

.77 × 268 = 206.

The same adjustments are made to obtain expected frequencies for the "Other" column. For men this would be

.23 × 246 = 57,

and for women,

.23 × 268 = 62.

respondents (independent of sex) who favor the treaty. To do this, we add the number of men who favor the treaty to the number of women who support it.

177 + 220 = 397

So we can say that 397 out of 514 people interviewed favored the treaty. This gives a proportion of .77, (that is, 397/514 = .77). Similarly, we can find the proportion of respondents who provided other responses. There were 69 men and 48 women who did not favor the treaty. This means that 117 of 514 respondents, or a proportion of .23, of the respondents did not favor the treaty.

3. Summarize the expected frequencies in a table.

	FAVOR	OTHER	TOTAL
MEN	189	57	246
WOMEN	206	62	268

4. Calculate the summary table exactly as you did for Box 12.1.

	MEN FAVOR	WOMEN FAVOR	MEN OTHER	WOMEN OTHER	TOTAL
1. O	177	220	69	48	514
2. E	189	206	57	62	514
3. $O - E$	−12	14	12	−14	0
4. $(O - E)^2$	144	196	144	196	680
5. $\dfrac{(O - E)^2}{E}$	.76	.95	2.53	3.16	7.40

5. Obtain $df = (\text{rows} - 1)(\text{columns} - 1) = 1 \times 1 = 1$.

$$\chi^2 = \frac{\Sigma (O - E)^2}{E}$$

Critical value at 1% significance level = 6.63
$$p < .01$$

Conclusion: Reject null hypothesis. Significantly more women than men favor a treaty freezing testing and deployment of nuclear weapons.

Once these proportions have been obtained, we can calculate the expected cell frequencies in the following manner. First let's consider men favoring the treaty. Under the null hypothesis .77 of the respondents favor the treaty. Apply this to the 246 men in the sample.

.77 × 246 = 189

So the expected frequency for Men Favor is 189. Now do the same for women.

.77 × 268 = 206

In other words, the expected frequency for Women Favor is 206. Similarly, we can obtain the same expected frequencies for the other responses. Remember, a proportion of .23 of the respondents (averaged across sex under the null hypothesis) gave a response other than favoring the treaty. For men, the expected frequency is

$$.23 \times 246 = 57,$$

and for women the expected frequency is

$$.23 \times 268 = 62.$$

In these calculations the expected values are always rounded to the nearest whole person. The expected frequencies for each cell in the table are summarized as Step 3 in Box 12.2.

Now we are ready to calculate the chi-square statistic. The final portion of Box 12.2 provides the summary table. You will notice that this table is very similar to the one presented in Table 12.2. The first row of the table gives the observed frequencies for each of the four cells in the table. The second row gives the expected frequencies. The third row describes the observed minus the expected frequencies for each cell. The fourth row is $(O - E)^2$. Now look at the final column in the summary table. Note that the sum of the expected and the sum of the observed frequencies are equal to one another. Also note that the sum of the observed minus the expected values is 0. This is exactly the same as it has been for previous problems. Again, these are conditions which must be met for the proper application of the chi-square statistic. Now look at the value at the end of the box. This is the chi-square statistic. In this example, chi-square equals 7.40. To interpret the chi-square value we need to find the number of degrees of freedom. For a chi-square contingency table, the number of degrees of freedom is

$$df = (\text{Rows minus } 1)\,(\text{Columns minus } 1).$$

In this example (step 3), there are two rows and two columns. So

$$df = (2 - 1)\,(2 - 1)$$
$$= (1)\,(1) = 1.$$

Now we must determine whether the chi-square value is statistically significant. This can be accomplished by examining the table in Appendix 5. Since there is only 1 degree of freedom, enter the table in the row for df = 1. Let's assume that we set the significance level for this test at .01, or the 1% level. According to the table, the critical value for chi-square at the 1% level with 1 degree of freedom is 6.63. Our obtained value of 7.40 exceeds this critical value. Therefore, we reject the null hypothesis and conclude that the probability is less than 1 in 100 that the observed difference between men and women in their attitude toward the nuclear treaty would have occurred by chance alone, that is, the differences between men and women are significant.

There is an alternative method for calculating chi-square in a 2 × 2 contingency table. In practice this method is probably easier than the one just described. The formula for calculating chi-square in a 2 × 2 table is

$$\chi^2 = \frac{N(AD - BC)^2}{(A + B)(C + D)(A + C)(B + D)},$$ (12.3)

where A, B, C, and D are cells. An example of the calculation of chi-square using this formula is given in Box 12.3. The example compares the San Diego Chargers and the Los Angeles Raiders for wins and losses during the 1983 NFL season. The research question is whether the distribution of wins and losses is the same or different for the two teams. In other words, is winning or losing independent of the particular teams? The null hypothesis is

H_0: Frequencies are independent,

and the alternative hypothesis is

H_1: Frequencies are not independent.

The first table in the box summarizes the win-loss records for the two teams. When working through this example, it will be important to learn some terms which statisticians frequently use. The cells of a 2 × 2 contingency table are usually labeled A, B, C, D. The location of these four cells is shown in the example. To use the formula, it will be necessary to obtain the *marginal sums*. For example, the marginal sum $A + B$ is the sum of cell A and cell B. It will be worthwhile to work through the steps presented in Box 12.3 to see if you can get the correct answer. It may also be valuable to calculate this same chi-square problem using the method presented above.

After working through the steps in Box 12.3, you should get a chi-square value of 4.57. As in the previous example there is 1 degree of freedom. The chi-square value of 4.57 exceeds the critical value for chi-square with 1 degree of freedom at the .05, or 5%, level. Therefore, we reject the null hypothesis and conclude that winning or losing is not independent of team for these two participants in the NFL. It may be appropriate to conclude that the Los Angeles Raiders won significantly more often than the San Diego Chargers.

One of the reasons this example was chosen is that it violates one of the important assumptions of the chi-square test. Do you know what it is? The answer is that the assumption of independence is violated. The San Diego Chargers and the Los Angeles Raiders are members of the same division in the National Football League. They play one another twice during the season. In other words, among the 16 games played by the San Diego Chargers, 2 of them are against the Los Angeles Raiders. For those 2 games, a win by one team is not independent of a loss by the other team. Technically, the chi-square test is not appropriate for examples such as this. Yet, in practice, chi-square is sometimes used to analyze data of this type. However, it should be used with great caution. The most extreme example would be one in which events are entirely dependent. For example, if the Raiders and the Chargers played one another each game of the season, then events in the table would be highly dependent.

● BOX 12.3

Step-by-Step Calculation of the 2 × 2 Chi-Square by Alternative Formula

H_0: The frequency of winning is independent of team
H_1: The frequency of winning is not independent of team

Formula:
$$\chi^2 = \frac{N(AD - BC)^2}{(A + B)(C + D)(A + C)(B + D)}$$

Significance or alpha level = .05

	WIN	LOSS	TOTAL GAMES
San Diego Chargers	6	10	16
Los Angeles Raiders	12	4	16
Total	18	14	32

Cell Labels:

A	B	A + B
C	D	C + D
A + C	B + D	N

STEPS

1. Find AD.

   $6 \times 4 = 24$

2. Find BC.

   $10 \times 12 = 120$

3. Obtain $AD - BC$.

   $24 - 120 = -96$

4. Square Step 3 to get $(AD - BC)^2$.

   $(-96)^2 = 9216$

5. Multiply Step 4 by N to obtain $N(AD - BC)^2$.

   $32 \times 9216 = 294{,}912$

6. Find each marginal sum. There are four different ones: $A + B$, $C + D$, $A + C$, and $B + C$.

   $6 + 10 = 16$ $6 + 12 = 18$
   $12 + 4 = 16$ $10 + 4 = 14$

7. Calculate the product of the marginal sums, or

   $(A + B)(C + D)(A + C)(B + D)$.
   $\quad 16 \ \times \ 16 \ \times \ 18 \ \times \ 14 \ = 64{,}512.$

8. Find χ^2 by dividing Step 5 by Step 7.

   $\dfrac{294{,}912}{64{,}512} = 4.57$

9. Determine df.

   df = (Rows − 1)(Columns − 1)
   $\quad = \quad 1 \quad \times \quad 1$
   $\quad = 1$

10. Look up critical value of χ^2 for 1 degree of freedom

   for 5% level critical value = 3.84.

11. Interpret the result.

   $4.57 > 3.84$

   Therefore, we reject the null hypothesis. The Los Angeles Raiders won significantly more often.

Other Nonparametric and Distribution-Free Procedures

Most of the statistical procedures that have been presented in previous chapters estimate one or more population parameters (such as standard deviation) on the basis of sample data. Using these methods requires that we make assumptions about population characteristics. For example, one of these assumptions is that the variable under study is normally distributed within the population from which the sample is drawn. For most applications of basic statistics, violations of this assumption have relatively little impact on a statistical test. Yet there are circumstances in which violations of assumptions about population distribution will cause substantial bias.

Statistical methods that estimate population parameters, such as the standard deviation, on the basis of sample data, are called **parametric statistics**. They are given this name because they are used to estimate the population parameters. There is another family of statistical procedures that does not make assumptions about population distributions. These procedures are called **nonparametric**, or distribution-free techniques. Many research workers prefer nonparametric methods because they rest on many fewer assumptions. These methods are also appropriate for other types of data. For example, nonparametric statistics are very valuable for the analysis of ordinal or rank data. Although some statisticians make the distinction between nonparametric and distribution-free methods (see Marasculio and McSweeney, 1977) the terms are usually used as synonyms. Nonparametric statistics should in itself be the topic of a detailed advanced statistics course. Many nonparametric methods enjoy common application in contemporary research. This chapter presents only a small sample of the available nonparametric and distribution-free procedures. The interested reader should consult Marasculio and McSweeney (1977) for a more exhaustive description of the available methods.

One of the advantages of nonparametric methods is that they are easy to use and, as we already noted, can be applied with less concern about underlying assumptions. Three nonparametric methods are presented in this chapter: the median test, the Mann–Whitney U-test, and the Kruskal-Wallis test.

■ The Median Test

The **median test** is one of the easiest statistical tests to apply. The purpose of the median test is to compare two groups, just as might be done with the t-test. However, with the median test, we evaluate the hypothesis that the two groups are sampled from a population with the same *median*.

First, let's review the conceptual basis for the median test. Suppose that we

• BOX 12.4

Median Test for Two Independent Samples

Information Processing Experiment—Unequal N

Assumptions: t-test not appropriate because of skewed distributions. Significance level .05.

H_0: $Md_1 = Md_2$
H_1: $Md_1 \neq Md_2$

Raw data:

	GROUP	
	Elderly Women	Daughters
	3	1
	4	0
	6	4
	7	5
ERROR	5	6
SCORE	26	0
	9	0
	21	7
	11	12
	14	6
		5
		4

STEPS

1. Make a frequency distribution of the data pooling the scores for both groups.

SCORE	f	SCORE	f
26	1	6	3
21	1	5	3
14	1	4	3
12	1	3	1
11	1	1	1
9	1	0	3
7	2		

2. Find the depth of median and the median for the distribution.

$$\frac{N+1}{2} = \frac{23}{2}$$
$$= 11.5$$
Median = 5.5

3. Tabulate the number of scores above the median and below the median in each group.

have two groups of subjects. The median can be calculated for each group or a single median can be calculated using scores from both groups combined. If the two groups are sampled from a population with the same median, we would expect half of the subjects in each group to have scores at or above the median and half to have scores below the median. If the two groups are sampled from different populations, the common median should not divide each group into equal halves. Thus, the logic of the median test is to determine whether a common median splits each of two groups into two equal subgroups.

The median test is illustrated in Box 12.4. This example is taken from a study of information processing in elderly women. The investigator wanted to know whether a certain aspect of memory declines as a function of age. She

GROUP	AT OR ABOVE MEDIAN	BELOW MEDIAN	TOTAL
Elderly Women	7	3	10
Daughters	4	8	12

4. Calculate the chi-square statistic using the method presented in Box 12.3. (See Box 12.3 if you need the step-by-step instructions.)

$$\chi^2 = \frac{N(AD - BC)^2}{(A + B)(C + D)(A + C)(B + D)} \quad (12.3)$$

	AT OR ABOVE MEDIAN	BELOW MEDIAN	
Elderly Women	A 7	B 3	$A + B$ 10
Daughters	C 4	D 8	$C + D$ 12
	$A + C$ 11	$B + D$ 11	

$$\chi^2 = \frac{22[(7 \times 8) - (4 \times 3)]^2}{(10)(12)(11)(11)}$$

$$= \frac{42592}{14520}$$

$$= 2.93$$

5. Find the degrees of freedom for the chi-square value.

$$df = (\text{rows} - 1)(\text{columns} - 1)$$
$$= (2 - 1)(2 - 1)$$
$$= (1)(1)$$
$$= 1$$

6. Look up the critical chi square value for 1 degree of freedom in Appendix 5.

Critical value for $df = 1$ (.05) $= 3.84$

$$2.93 < 3.84$$

Differences not significant; retain null hypothesis.

obtained a sample of elderly women (ages 65–75) and gave them a memory test. As a control group, she found the daughters of the elderly subjects. The daughters were all 25–35 years younger than their mothers. This design was chosen to control for genetic differences in memory, social class, and sex. Two of the elderly subjects had two daughters who participated in the study. Therefore, the number of controls was larger than the number of elderly subjects by 2. The median test can be used with equal or unequal sample sizes. Therefore, the unequal N poses no problem. The investigator chose the median test because she was concerned that memory scores were not normally distributed in the elderly population. Previous research had shown that a few elderly subjects have severely impaired memories and do very poorly on these tests. Thus, the population distribution was suspected to be severely skewed.

The null hypothesis for medians Md_1 and Md_2 is

H_0: $Md_1 = Md_2$.

This is tested against the alternative hypothesis

H_1: $Md_1 \neq Md_2$.

Calculating the median test requires several steps. First you must find the median for the two groups combined using the method presented in Chapter 2. In this example, the median, or the score marking the 50th percentile in the distribution, is 5.5.

The next step is done separately for each of the two groups. The number of subjects in each group who fall at or above the median and below the median is tabulated. In this example, the score on the memory test is the number of errors. Thus, higher scores represent more errors. In the group of elderly women, 7 obtained error scores at or above the median for the combined groups while 3 obtained scores below the median. In the sample of daughters, 4 obtained scores at or above the median while 8 obtained scores below the median.

If you look at Step 3 in Box 12.4 you will find a 2×2 contingency table. This can be analyzed using the 2×2 chi-square method presented in the last section. Step 4 in Box 12.4 illustrates how this is done. In summary, the 2×2 chi-square analysis is used as the analysis step in the median test. In this example the value of chi-square is 2.93. There is 1 degree of freedom. According to Appendix 5 the critical value for chi-square with 1 degree of freedom is 3.84. Since 2.93 is less than 3.84, we conclude that the differences between these two groups are not statistically significant. In other words, we do not reject the null hypothesis that the two groups were sampled from a population with the same median.

The conclusion reached in this analysis may be somewhat surprising to you. Simply looking at the data gives the impression that these two groups are indeed quite different. One of the difficulties with nonparametric statistics is that they have less *power* to detect significant differences between groups. This point is discussed in more detail later in the chapter.

■ The Mann–Whitney *U*-Test

The **Mann–Whitney *U*-test** is one of the most popular nonparametric procedures. This test is the nonparametric counterpart of the *t*-test. The Mann–Whitney *U*-test is an analysis of ranks for rank data. It is used instead of the *t*-test when assumptions about population distributions might be violated or when sample sizes are very small. Theoretically, the Mann–Whitney *U*-test is a test of the difference between two population distributions rather than a test of the difference between two means or medians. The null hypothesis is that a single, combined population distribution can represent the two sample population distributions. The null hypothesis is likely to be supported if the two sample distributions greatly overlap and likely to be rejected if the two population distributions are separate from one another. This is illustrated in Fig. 12.1.

The Mann–Whitney U-test is based on the differences in ranks of scores between two groups. The scores for the two groups are rank ordered in a common pool. Then the sum of the ranks is obtained separately for each group. If the two distributions overlap greatly, then the sum of the ranks for the two groups will be very similar. On the other hand, if the sums of the ranks for the two groups are very different, we might conclude that a common distribution does not represent each group equally well. The Mann–Whitney U-test is used to formally evaluate differences between ranks.

An example of the Mann–Whitney U-test is described in Box 12.5. Box 12.6

Figure 12.1

The Mann–Whitney U-test is a test of population distributions.
Under the null hypothesis, a common distribution is adequate to represent both groups. This may be supported in the top example where the two sample distributions greatly overlap. However, this null hypothesis is less likely to be supported in the lower example in which the distributions have little overlap.

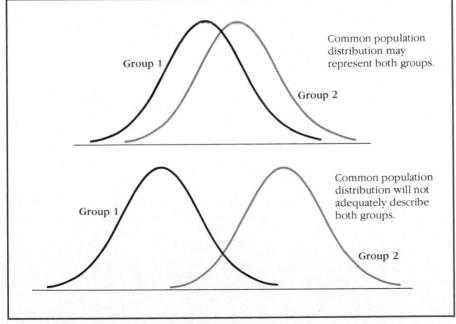

Group 1   Group 2

Common population distribution may represent both groups.

Group 1   Group 2

Common population distribution will not adequately describe both groups.

Source: After Minium and Clark, 1982.

● BOX 12.5

Step-by-Step Computation of Mann–Whitney U-Test

Example: Trials to criterion for rhesus monkey under drug and placebo conditions.

Formulas are given in Box 12.6.

H_0: A common distribution is adequate to represent both groups.

H_1: H_0 is false.

	GROUP 1 DRUG CONDITION SCORE	GROUP 2 PLACEBO CONDITION SCORE
TRIALS TO CRITERION	31	17
	46	18
	19	22
	26	29
	27	25
	31	16
	19	12
	42	10
	47	33
	32	14
	51	
	17	

Score	Rank	Score	Rank
51	1	25	12*
47	2	22	13*
46	3	19	14.5
42	4	19	14.5
33	5*	18	16*
32	6	17	17.5*
31	7.5	17	17.5
31	7.5	16	19*
29	9*	14	20*
27	10	12	21*
26	11	10	22*

2. Place the ranks on the raw data table next to each raw score.

DRUG CONDITION		PLACEBO	
Score	Rank	Score	Rank
31	7.5	17	17.5
46	3	18	16
19	14.5	22	13
26	11	29	9
27	10	25	12
31	7.5	16	19
19	14.5	12	21
42	4	10	22
47	2	33	5
32	6	14	20
51	1		
17	17.5		

STEPS

1. Make a distribution of scores, arranging them from highest to lowest. When two scores are tied, they take up two ranks and each gets the average of these two ranks. For example, there are two animals with a score of 19. These animals would be the 14th and 15th in the order. So, each would as assigned the rank 14.5. Identify scores in one group with a *, (in this case, the placebo condition).

3. Sum the ranks for Group 1 to get R_1.

$$R_1 = 7.5 + 3 + 14.5 + \cdots + 17.5 = 98.5$$

4. Sum the ranks for the second group to get R_2.

$R_2 = 17.5 + 16 + 13 + \cdots + 20 = 154.5$

5. Find $N_1 N_2$.

$N_1 N_2 = (12)(10)$
$= 120$

6. Calculate $N_1(N_1 + 1)$.

$N_1(N_1 + 1) = 12(12 + 1)$
$= 156$

7. Divide Step 6 by 2.

$\dfrac{N_1(N_1 + 1)}{2} = \dfrac{156}{2} = 78$

8. Calculate the U_1-statistic using the formula

$U_1 = N_1 N_2 + \dfrac{N_1(N_1 + 1)}{2} - R_1.$

This can be accomplished using the results of Steps 3, 5, and 7.

$U_1 = \text{Step 5} + \text{Step 7} - \text{Step 3}$
$= 120 + 78 - 98.5$
$= 99.5$

9. Obtain the expected value of U using the formula

$U_E = \dfrac{N_1 N_2}{2}, \quad \text{or} \quad \dfrac{\text{Step 5}}{2}.$

$= \dfrac{120}{2} = 60$

10. You now need to calculate the standard deviation of U. This will require one new quantity: $(N_1 + N_2 + 1)$. Find this.

$(N_1 + N_2 + 1) = (12 + 10 + 1) = 23$

11. Now you can calculate the standard deviation of U as

$\sigma_U = \sqrt{\dfrac{N_1 N_2 (N_1 + N_2 + 1)}{12}}$

$= \sqrt{\dfrac{(\text{Step 5})(\text{Step 10})}{12}}$

$= \sqrt{\dfrac{(120)(23)}{12}}$

$= \sqrt{230}$

$= 15.17.$

12. The final calculation is to convert the U-statistic into a Z-score. If you are fuzzy on Z-scores, review Table 4.1.

$Z = \dfrac{U_1 - U_E}{\sigma_U}$

$= \dfrac{\text{Step 8} - \text{Step 9}}{\text{Step 11}}$

$= \dfrac{99.5 - 60}{15.17}$

$= \dfrac{39.5}{15.17}$

$= 2.60$

13. The critical value is obtained from the Z or standard normal distribution when N for each sample is larger than 8. The Z-distribution is shown in Appendix 1. At the .05 level (two-tailed) the critical value is 1.96.

Since $2.60 \geq 1.96$ we reject the null hypothesis and conclude that the observed difference between distributions is statistically significant.

● BOX 12.6

Formulas for Mann–Whitney *U*-Test

Box 12.5 gives step-by-step instructions for calculating the Mann–Whitney *U*-test. Once you understand the routine, these formulas can be used to perform the test.

$$U_1 = N_1 N_2 + \frac{N_1(N_1 + 1)}{2} - R_1, \tag{12.4}$$

where U_1 is the *U*-value for Group 1 and R_1 is the sum of the ranks for Group 1.

$$U_E = \frac{N_1 N_2}{2} \tag{12.5}$$

$$\sigma_U = \frac{N_1 N_2 (N_1 + N_2 + 1)}{12} \tag{12.6}$$

$$Z = \frac{U_1 - U_E}{\sigma_U} \tag{12.7}$$

The test is evaluated using the *Z*-distribution.

gives the formulas for the calculation of the Mann–Whitney *U*-test. If you are comfortable using the formulas, you may wish to try the Mann–Whitney *U*-test without going through the step-by-step instructions provided in Box 12.5.

Box 12.5 presents data from an experiment on the effects of Valium (Diazepam) on learning for rhesus monkeys. The animals were all experienced with the test apparatus. They were randomly assigned to perform a discrimination learning task one hour after an injection of 3 mg of Valium or 3 mg of a placebo. The dependent variable was the number of trials the monkey required to reach a criterion of six consecutive correct trials on a two-choice discrimination learning task. The higher the score, the poorer the performance.

The first step in the calculation of the Mann–Whitney *U*-test is to list the raw data for the two groups. *If one group has more subjects than the other, that must be designated as Group 1.* In this experiment the drug condition had 12 subjects while the placebo condition had only 10. Thus, the drug condition was designated as Group 1. Next make the common distribution of scores by arranging the raw data from highest to lowest. (Actually, with the Mann–Whitney *U*-test it does not matter whether the scores are arranged in ascending or descending order; for this example we have chosen to arrange them in descending order.) Once the scores have been arranged, a rank can be assigned to each.

One common problem is ties between scores. Very often, two or more subjects will get exactly the same score. If two scores are the same, assign two ranks to them and then assign each score the average of these two ranks. For instance, in this example two monkeys each made 31 errors. Notice in Box 12.5 that the monkeys with the score of 31 are the seventh and eighth in the distribution. To determine their rank, take the average of these two ranks:

$$\frac{(7 + 8)}{2} = 7.5.$$

If three subjects have the same score, assign three different ranks to them and take the average of the three. For example, if three subjects had the same score and they appear as the tenth, eleventh, and twelfth listed in the arrangement of scores, take the average of these three ranks:

$$\frac{(10 + 11 + 12)}{3} = 11$$

and assign that number to each subject.

When arranging the scores in rank order, it is often convenient to place asterisks by those in one group so they are easily identified. Then make a list of the rank orders for each of the two groups. The next step requires you to take the sum of the ranks for Group 1. This is shown in Box 12.5 as Step 3. Steps 5 through 8 in Box 12.5 display the calculation of the U-statistic for Group 1.

The Mann–Whitney U-test also requires the calculation of an expected value for the U-statistic. The expected value is simply the product of the sample sizes divided by 2, or

$$U_E = \frac{N_1 N_2}{2}. \tag{12.5}$$

This is shown as Step 9 in Box 12.5. With the expected value for the U, it is possible to find the difference between the observed and the expected values. However, it is also important to determine whether this difference between expected and observed is greater than would be expected by chance. Doing so requires a standard deviation score. The standard deviation of U can be calculated as shown in Step 11. With the observed and the expected U-values and the standard deviation, it is possible to calculate a Z-score for the U-statistic. This is shown in Step 12. In this example, the Z-score is 2.60. The standard normal distribution Z-scores were discussed in Chapter 4. You might recall that the critical value in the Z-distribution for the .05 significance level (two-tailed test) is 1.96. Since the observed value of 2.60 exceeds 1.96, we can reject the null hypothesis and conclude that the differences between the two groups are statistically significant. It appears that Valium hinders performance (high scores mean poorer performance) in this group of monkeys.

For small samples there is a shortcut for the evaluation of the Mann–Whitney U-test. This shortcut applies when neither group exceeds 15 subjects. This method requires only that you find the sum of the ranks for Group 2. As shown in Box 12.5, the sum of the ranks for Group 2 is 154.5. This information is evaluated using the table in Appendix 6. This is a special table of critical values for the sum of ranks in the Mann–Whitney test. To evaluate the rank, find the portion of the table for $n_1 = 10$. Then find the row for $n_2 = 12$. The entries in the table are the critical values at different probability levels. Go across the row for $n_2 = 12$ until you come to the column labeled .025. This is for the 5% significance level in a two-tailed test ($.05/2 = .025$). The numbers listed there are 84 and 146. These define the lower and upper boundaries for the 95% confidence interval. Since 154.5 is larger than 146 (the upper limit) we reject the null hypothesis and conclude that the differences between groups are statistically significant.

One final note on the Mann–Whitney U-test. For the calculation of the Z-score, you can use either U_1 or U_2. If you had used U_2, you would use the following formula

$$U_2 = N_1 N_2 + \frac{N_2(N_2 + 1)}{2} - R_2. \qquad (12.8)$$

R_2 was calculated as Step 4 in Box 12.5. The rest of the calculations follow.

$$N_1 N_2 = 12 \times 10 = 120$$
$$N_2(N_2 + 1) = 10(10 + 1) = 110$$
$$R_2 = 154.5 \quad \text{(from Box 12.5)}$$
$$U_2 = 120 + \frac{110}{2} - 154.5 = 20.5 \qquad (12.9)$$
$$Z = \frac{U_2 - U_E}{\sigma_U} = \frac{20.5 - 60}{15.17} = -2.60$$

Notice that the Z-value is exactly the same as that obtained in Box 12.5 except that it has the opposite sign. All problems will come out this way; this can be used as a check on the accuracy of your calculations when using the Mann–Whitney U-test.

You should be cautious when using the Mann–Whitney U-test if there are a large number of tied ranks. When many scores are the same, the test will be less meaningful. As we have shown, special tables can be used to evaluate summed ranks when sample size for each group is 15 or less. The transformation of the U to Z is appropriate when each of the samples is of size 8 or larger.

■ The Kruskal–Wallis Test

The Mann–Whitney U-test is the nonparametric counterpart of the t-test. The Kruskal–Wallis test is an extension of the Mann–Whitney U-test for three or more groups. This test might be the considered the nonparametric counterpart of the one-way analysis of variance. The rationale for the Kruskal–Wallis test is very much the same as that for the Mann–Whitney U-test. The purpose of the test is to evaluate three or more sampling distributions of ranked data. The null hypothesis is that each sample distribution is represented equally well by a single distribution of ranks.

The calculation of the Kruskal–Wallis test is shown in Box 12.7. Students not wishing to go through the step-by-step calculation may benefit from studying the formulas presented in Box 12.8.

● BOX 12.7

Step-by-Step Calculation of the Kruskal–Wallis Test

Rating of employee performance for firemen in three shift schedules

$$H = \frac{12}{N(N+1)} \sum \left(\frac{R_i^2}{n_i} \right) - 3(N+1) \qquad (12.10)$$

H_0: Each distribution of ranks is represented equally well by a common distribution of ranks.

H_1: H_0 is false.

Ranked Data

R_1 GROUP 1 (DAY)	R_2 GROUP 2 (NIGHT)	R_3 GROUP 3 (DAY/NIGHT)
1	2	4
3	7	8
5	9	11
6	12	13
10	14	$R_3 = 36$
$R_1 = 25$	$R_2 = 44$	

STEPS

1. Combine the scores into a single distribution and rank order them. The lowest score should be given the rank 1, the second lowest rank 2, and so on. In this example the ratings are provided as ranks (that is, the data are already in a single ranked distribution).

2. Sum the ranks separately for each group (R_i).

$R_1 = 1 + 3 + 5 + 6 + 10 = 25$
$R_2 = 2 + 7 + 9 + 12 + 14 = 44$
$R_3 = 4 + 8 + 11 + 13 = 36$

3. Square each sum of ranks (R_i^2).

$R_1^2 = 25^2 = 625$
$R_2^2 = 44^2 = 1936$
$R_3^2 = 36^2 = 1296$

4. Divide each squared group rank by its own sample size (R_i^2/n_i).

$\dfrac{R_1^2}{n_1} = \dfrac{625}{5} = 125$

$\dfrac{R_2^2}{n_2} = \dfrac{1936}{5} = 387.2$

$\dfrac{R_3^2}{n_3} = \dfrac{1296}{4} = 324$ *(continued)*

• BOX 12.7 (*continued*)

5. Sum the values obtained in Step 4.

$$\sum \frac{R_i^2}{n_i} = 125 + 387.2 + 324 = 836.2$$

6. Calculate the $12/N(N + 1)$ portion of the formula.

$$\frac{12}{N(N + 1)} = \frac{12}{14(14 + 1)}$$

$$= \frac{12}{(14)(15)}$$

$$= \frac{12}{210}$$

$$= .057$$

Note: Serious errors can be made by rounding this step. Be sure to carry your calculation to at least 3 significant decimals.

7. Find

$$3(N + 1) = 3(14 + 1)$$

$$= 3(15)$$

$$= 45.$$

8. Calculate

$$H = \frac{12}{N(N + 1)} \sum \frac{R_i^2}{n_i} - 3(N + 1)$$

using Steps 5, 6, and 7.

$$H = [(\text{Step 6})(\text{Step 5})] - \text{Step 7}$$

$$= [(.057)(836.2)] - 45$$

$$= 47.66 - 45$$

$$= 2.66$$

9. The statistical significance of H can be evaluated using the chi-square distribution. The degrees of freedom are

$$df = \text{groups} - 1 = 3 - 1$$

$$= 2.$$

10. Look up the critical χ^2 value for the number of degrees of freedom.

This example has $df = 2$ and uses .05 significance level. According to Appendix 5 the critical value is 5.99. Since $2.66 < 5.99$, the differences between groups are considered nonsignificant.

The example shown in Box 12.7 is taken from supervisor ratings of firemen. A group of 14 firemen in a small city was randomly divided into three groups. For a period of one month, one group was assigned to duty only during daylight hours (9:00 AM to 5:00 PM). A second group was assigned only nighttime hours, while the third group worked half days and half nights. After one month, all of the participating firemen were assigned to the same daytime shift for one week. Then they were evaluated by the chief, who rank ordered the men according to their performance. The department was interested in knowing whether the different working hours had an impact on the firemen's performance.

The first step in the calculation of the Kruskal–Wallis test requires that all cases be combined into a common group and rank ordered. In this example the data were provided as ranks; therefore, we can go right to Step 2. (If, however, the data had not been in the form of ranks, they would be combined and rank ordered with the lowest score being assigned rank 1, the second lowest score assigned rank 2, and so on, until the highest score was assigned the

• BOX 12.8

Formula of Kruskal–Wallis Test

The Kruskal–Wallis test is used to compare two or more groups. It does not depend on the restrictive assumptions of the analysis of variance. Step-by-step instruction for the calculation of the Kruskal–Wallis test is given in Box 12.7. The test can also be calculated from this formula.

Formula:
$$H = \frac{12}{N(N+1)}\left[\sum \frac{R_i^2}{n_i}\right] - 3(N+1),$$
(12.10)

where

N = total number of cases,
n_i = the number of cases in a particular group, and
R_i = the sum of the ranks in a particular group.

The H-statistic is evaluated using the chi-square distribution with

df = number of groups − 1.

highest rank.) Ties would be handled the same way as they were for the example in the Mann–Whitney U-test.

Once ranks have been assigned, the ranks within each group are summed. In this example, the sum of the ranks for Group 1 (daytime hours) was 25; it was 44 and 36 for the other two groups, respectively. Next these sums of group rankings are each squared and divided by the number of cases within that group. The Kruskal–Wallis test can be performed for unequal sample sizes. However, bias may result when the sample sizes are very unequal. Steps 4 through 8 in Box 12.7 demonstrate the calculation of the H-statistic. Unlike many of the other tests described in this book, there is no appendix for the H-distribution. Instead, the calculated value for H can be evaluated using the chi-square distribution.

To determine the statistical significance of H, we must first find the degrees of freedom. For the Kruskal–Wallis test, the degrees of freedom are equal to the number of groups minus 1.

df = Groups − 1.

So in this example, $df = 3 - 1 = 2$. For this study, the investigator decided to set the significance level at .05. To reject the null hypothesis, the observed value of H must exceed the tabled chi-square value for the .05 significance level with 2 degrees of freedom. According to Appendix 5, this value is 5.99.

The observed H-value of 2.66 falls below 5.99. Therefore, the null hypothesis is retained, and differences among ranks for the three groups are attributed to chance.

The assumptions for the Kruskal–Wallis test are very similar to those for the Mann–Whitney U-test. The test assumes that cases are randomly sampled and that there are no ties. However, in practice, ties in ranks produce relatively little bias in the use of the test.

Sometimes, there is a reason to perform the Kruskal–Wallis test when the sample size for each group is very small. Special tables are available to evaluate the significance of H when there are 5 or fewer cases in each group. In practice, experimental effects would have to be very large to detect an effect with only 5 cases per group. In the next section, some practical issues relevant to sample size will be discussed.

Power Problems

Nonparametric or distribution-free methods have several advantages. They are easy to use, and they rest on very few assumptions about population distributions. Thus, they are attractive for many purposes.

Despite the advantages of nonparametric procedures, there are also some disadvantages. The most notable is that they are less powerful. The power of statistical tests was discussed in Chapter 7. Recall that the *power of the test* is the probability of rejecting the null hypothesis when indeed it should have been rejected. When a test has low power, the researcher may accept the null hypothesis even though it should have been rejected. In other words, the researcher may conclude that there were no differences among treatments or groups even though these groups really did differ. The difficulty was that the statistical test was unable to detect the differences.

Given the same sample size, nonparametric statistics typically have lower power than their parametric counterparts. Some statisticians compare statistical tests according to their **relative efficiency**. The relative efficiency of two statistical tests is the ratio of their sample sizes required to obtain the same level of statistical power. Nonparametric and distribution-free methods tend to be relatively less efficient than their parametric counterparts. Therefore, most researchers prefer parametric tests when they are appropriate. These parametric tests rest on many assumptions such as the normal distribution of the population from which samples are drawn. Minor violations of these assumptions have relatively little effect upon the validity of the statistical test. However, when there are serious violations of these assumptions, nonparametric statistics are appropriate.

Summary

Chi-square statistics are used to evaluate frequency data. A common use of the chi-square test is to compare frequencies across different categories. Chi-square can also be used to evaluate bivariate frequency distributions obtained from the combination of two or more variables. The 2 × 2 contingency table is commonly used to compare two groups on a nominal dependent variable with two categories.

Most statistical tests for comparing the means of two or more groups depend on well-specified assumptions. For example, the t-test and the analysis of variance assume that scores are sampled from a population with a normal distribution and that the variance within each group is equal. With large samples, violations of these assumptions do not always severely bias the test. However, there are some cases in which the violation of these assumptions can produce serious biases. Two cases are when sample sizes are very small and when the population distribution is severely skewed. Under these circumstances it is more appropriate to use nonparametric, or distribution-free, methods. These methods are based on fewer assumptions about the population characteristics.

Three nonparametric methods were presented. The first is the median test, which is a simple procedure for comparing two groups. Cases across the two groups are combined and a common median is calculated. Then the number of cases within each group falling at or above the median and below the median is tabulated. This process results in a 2 × 2 contingency table which is evaluated using the chi-square statistic.

The Mann–Whitney U-test is another commonly used nonparametric statistic. This test requires placement of all cases in a combined rank order. The test evaluates differences in ranks to determine whether a common ranking distribution represents each of two groups equally well. The Mann–Whitney is considered a nonparametric counterpart for the t-test since it is used to compare two independent groups. The Kruskal–Wallis test is very similar to the Mann–Whitney U-test except that it is used to evaluate three or more groups. Therefore, the Kruskal–Wallis test is considered a nonparametric counterpart for the one-way analysis of variance. There are many different nonparametric procedures. This chapter has only provided an overview of some of the most commonly used methods.

Although there are many advantages to nonparametric statistics, there are also some disadvantages. In particular, nonparametric statistics typically have lower statistical power than their parametric counterparts. As a result, nonparametric statistics are considered to be relatively less efficient because they require more subjects to achieve the same level of statistical power than do parametric methods.

Key Terms

1. **Frequency Data**   Raw data showing the frequency for observations in each category in a nominal scale.
2. **Chi-Square Test**   A statistical test used to evaluate the null hypothesis when the dependent variable is expressed as a frequency. The test is based on a comparison of expected to observed frequencies. Specifically it is

    $$\chi^2 = \sum \frac{(O - E)^2}{E}.$$ (12.1)

3. **Parametric Statistics**   Statistical methods that estimate population parameters, such as the standard deviation, on the basis of sample data. These population parameters can also be differences between means such as those studied with the t-test or the analysis of variance.
4. **Nonparametric Statistics**   A family of statistical procedures that do not make assumptions about population distributions. This term is often used as a synonym for distribution-free statistics.
5. **Median Test**   A nonparametric statistical test that uses the chi-square method to evaluate the proportion of cases in each of two groups that fall at or above the median for both groups combined.
6. **Mann–Whitney U-Test**   A nonparametric method used to compare two groups. It is used when the parametric assumptions underlying the t-test cannot be used.
7. **Kruskal–Wallis Test**   A nonparametric method used to compare three or more groups. It is used when the assumptions of normality underlying the one-way analysis of variance cannot be met.
8. **Relative Efficiency of a Test**   The relative efficiency of two statistical tests is the ratio of sample sizes required to obtain the same level of statistical power. An efficient test has high power with low sample size while an inefficient test requires more subjects to obtain the same level of statistical power.

Exercises

1. One hundred scientists attend a national meeting. You are to determine whether four regions of the country are equally represented. Forty-three of the scientists are from the East, 19 are from the South, 21 are from the Midwest, and 17 are from the West. Are the four regions equally represented at the meeting?

2. A law school wants to determine whether male or female students have a
    higher probability of passing the bar examination in a certain state. Among
    82 male graduates in the class of 1986, 40 passed the exam and 42 failed.
    Among 69 female graduates, 47 passed the exam and 22 failed. How would
    you evaluate this question? State the null hypothesis, the alternative hy-
    pothesis, and then evaluate the question.

3. The following is a list of points scored by members of two basketball
    teams.

Team A	Team B
6	8
11	6
14	0
3	12
4	6
7	11
16	2
19	10

Use the median test to determine whether these two teams were sampled
from a population with the same median.

4. Using the data from Exercise 3, perform the Mann–Whitney U-test. State
    the null hypothesis, the alternative hypothesis, and then evaluate the ques-
    tion.

5. A group of patients with a sleep disorder were randomly assigned to take
    one of two different drugs or a placebo. Then their doctor observed them
    in a sleep laboratory and recorded the number of minutes it took them to
    fall to sleep. The results were as follows.

Drug 1	Drug 2	Placebo
0	1	6
2	3	21
4	4	17
1	6	9
0	4	2
3	2	1
7	8	16

Evaluate the results of the experiment using the Kruskal–Wallis test.

6. When does a test have high relative efficiency?

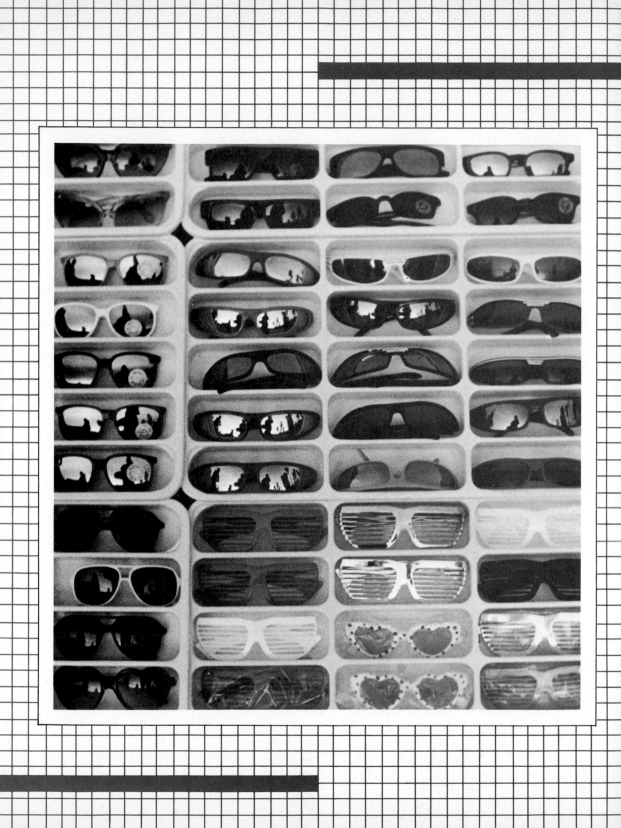

13

An Overview of Statistics

You have now come to the end of this basic statistics book. Many courses in the behavioral and health sciences require pyramidlike learning. Pyramid learning requires that a firm foundation be placed before the next step can be undertaken. Statistics requires this sort of pyramid learning perhaps more so than any other area in the behavioral and health sciences. In some courses, it is possible to survive even though the previous material was not well mastered. This is not so with statistics. Without an understanding of the concepts presented in this book you will have difficulty in the more advanced courses. Thus, it is important to go back and review introductory statistics in preparation for your next statistics course.

A Review of Some Basic Concepts

The first portion of this book was devoted to descriptive statistics. In Chapter 2 a review of scales of measurement was presented. There are four basic levels of measurement: nominal, ordinal, interval, and ratio. Knowing something about the scale of measurement is extremely important for selecting the most appropriate statistical procedure. A nominal scale of measurement is considered to be *unscaled*. Interval and ratio data are considered to be *scaled*, and ordinal data are usually treated as though they are scaled. Some statistical tests such as chi-square are appropriate for nominal data. Others, such as the Mann–Whitney U-test and the Kruskal–Wallis test, are designed for ordinal data. Scaled scores such as interval or ratio scores are most appropriately analyzed using methods such as the analysis of variance.

Independent variables should also be evaluated for being scaled or unscaled. Most of the statistical procedures presented in this book assume that the independent variable is nominal or unscaled. For example, the independent variable in a t-test is the group, that is, treatment, control, and so forth. There are some circumstances in which the independent variable is scaled. Under these circumstances methods such as correlation and regression are most appropriate. No matter how advanced your statistical skills become, you will always find yourself referring back to basic principles.

Putting Together the Many Things You Have Learned

Many of the methods you have mastered in this first course must serve as building blocks for more advanced pursuits in statistics. Indeed, many of the procedures presented in the latter chapters in this book are built on foundations provided in the earlier sections. Table 13.1 lists a few statistical procedures and shows some of the foundations they are built on. This is a very incomplete list and is only used to show you how some of these methods depend on common foundations. For example, to calculate the variance, you must first know the mean. To find the standard deviation, you must first know the variance. A Z-score cannot be calculated unless the mean and standard deviation are known first. Statistical tests such as the t-test require knowledge of a variety of concepts including the mean, standard devation, variance, and

Table 13.1	
A look at some statistical building blocks	
PROCEDURE	SELECTED BLOCKS
Variance	Mean
Standard deviation	Variance, mean
Z-score	Mean, standard deviation
Standard error of the mean	Mean, standard deviation, variance, N
t-test	Mean, standard deviation, variance, standard error of the mean, probability distributions
Analysis of variance	Mean, variance, sums of squares, mean squares, probability distributions
Chi-square	Frequency distributions, expected values, probability distributions
Mann–Whitney U-test	Z-scores, ranking methods, probability distributions
Median test	Chi-square

standard error of the mean, as well as an understanding of probability and probability distributions.

Although many statistical tests have been covered, there are some similarities among them. All of the statistical tests presented in this book are ratios. The numerator of the ratio is the observed statistic. The denominator of the ratio gives the expected value of the statistic. For example, with the t-test, the numerator is the observed difference between means for two groups. The denominator is the expected difference between these groups based on sampling variation. For each statistical test, the larger the value of the test statistic, the less likely that the observations are the result of chance alone. To interpret test statistics we must refer to probability distributions. Different statistical tests are referenced against different tables of known probabilities. However, the different statistical tests are all used to make statements about the probability that the observations could have occurred by chance alone.

Calculation and Computers

Throughout this book, you have encountered step-by-step methods for calculating the various statistical procedures. When first learning statistical methods it is important to calculate the procedures by hand. The step-by-step methods give you a feel for how to get the appropriate answer. As you progress, the step-by-step calculations may be unnecessary, and you will be able to make the appropriate calculations after simply examining the formula.

Within the last two decades there have been remarkable advances in computing. Now it may actually be unusual for a statistician to calculate a statistical test by hand. Hand-held calculators perform many of the procedures described in this book. In addition, software is available for both large and small computers to perform many of these calculations. The workbook provides descriptions for calculating statistical tests using microcomputers. Using these machines most students will be able to find the right answers to statistical problems. In the future, the real challenge will be to pick the right statistical method and to understand its proper interpretation.

Statistical Inference and Experimental Design

The correct calculation of statistical tests is only one small part of research work. One topic that is very close to statistics is experimental design. If an experiment is not designed properly, statistical methods will be of little value for rescuing a meaningful interpretation. Many statistical methods go hand in hand with experimental design. For example, the analysis of variance was designed for studies in which subjects are selected from a population and randomly assigned to experimental groups receiving different treatments. However, many people use the analysis of variance for experiments with other designs. There are many such examples in medical research. Consider an experiment on a new cancer drug. Since the drug has unknown effects, it is only given to those with no other hope for survival. A placebo is given to a group of patients with less severe cancer and to a control group that does not have cancer. Can there be a meaningful interpretation of this experiment?

This experiment has many problems. The most severe one is that at least two variables are being manipulated simultaneously. The first variable is severity of illness and the second is the medication. The group receiving a new treatment differed from the other two groups in at least two ways:

1. they received a new drug, and
2. they were more severely ill.

If there are more deaths in this group, it is difficult to separate the toxic effects of the drug from the poor prognosis for the patients.

When evaluating and planning research it is always important to determine whether the statistical tests have been performed properly. However, skillful use of statistical methods is only part of the process. You also must appraise the experimental design to determine whether the results are meaningful.

Choosing a Statistical Test For Different Research Designs

Table 13.2 presents a flow diagram showing how a particular statistical test might be selected for use in conjunction with different research problems. This table is also presented in the form of a "statistical consultant" program in conjunction with the workbook. This is presented only to give you some basic information about the selection of a statistical test. The selection of a statistical procedure is often complex and dependent on specific features of the research. Therefore, this table and the program cannot substitute for a real (human) statistical consultant.

Looking Ahead to More Advanced Courses

In more advanced courses you will be able to build on the skills obtained in this introductory course. These courses are typically available at the graduate

Table 13.2

Statistical consultant

1. Do you have one group of scores or more than one group of scores?

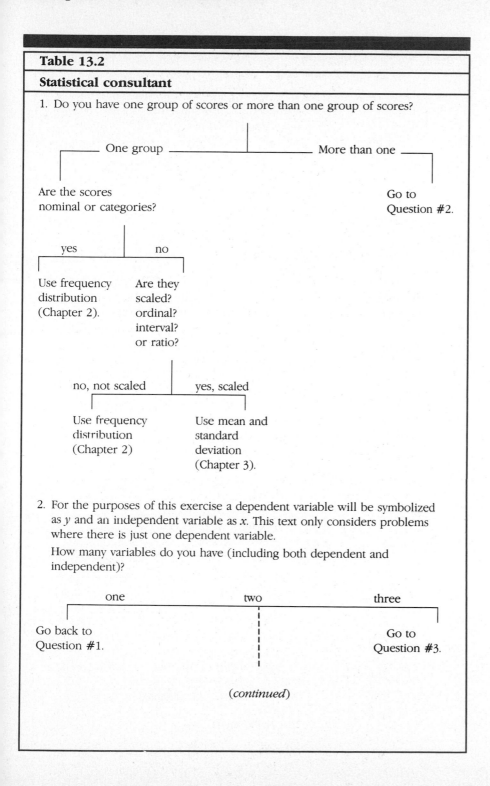

One group ———————— More than one

Are the scores
nominal or categories?

Go to
Question #2.

yes   no

Use frequency   Are they
distribution   scaled?
(Chapter 2).   ordinal?
                  interval?
                  or ratio?

no, not scaled   yes, scaled

Use frequency   Use mean and
distribution   standard
(Chapter 2)   deviation
                  (Chapter 3).

2. For the purposes of this exercise a dependent variable will be symbolized
as y and an independent variable as x. This text only considers problems
where there is just one dependent variable.

How many variables do you have (including both dependent and
independent)?

one   two   three

Go back to   Go to
Question #1.   Question #3.

(continued)

Table 13.2 (*continued*)

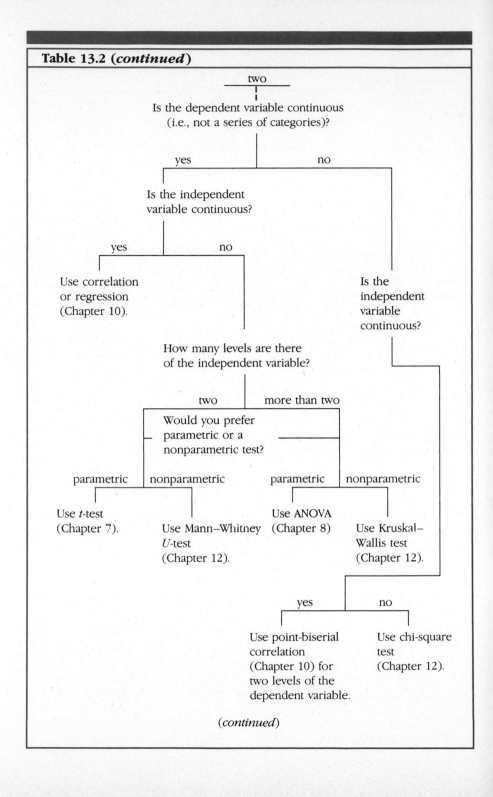

two

Is the dependent variable continuous
(i.e., not a series of categories)?

yes   no

Is the independent
variable continuous?

yes   no

Use correlation
or regression
(Chapter 10).

Is the
independent
variable
continuous?

How many levels are there
of the independent variable?

two   more than two

Would you prefer
parametric or a
nonparametric test?

parametric   nonparametric   parametric   nonparametric

Use *t*-test
(Chapter 7).

Use Mann–Whitney
U-test
(Chapter 12).

Use ANOVA
(Chapter 8)

Use Kruskal–
Wallis test
(Chapter 12).

yes   no

Use point-biserial
correlation
(Chapter 10) for
two levels of the
dependent variable.

Use chi-square
test
(Chapter 12).

(*continued*)

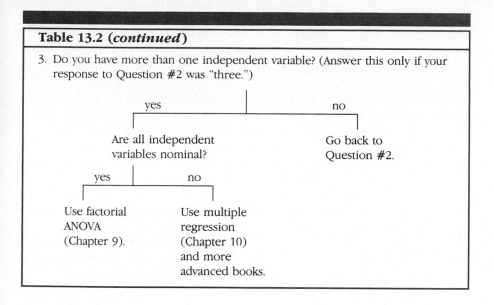

Table 13.2 (*continued*)

3. Do you have more than one independent variable? (Answer this only if your response to Question #2 was "three.")

level or in some cases the advanced undergraduate level. Following are very brief descriptions of courses that you may wish to pursue in the future.

EXPERIMENTAL DESIGN. This course typically is an extension of analysis of variance methods. In Chapter 9, you learned about the two-way analysis of variance. In an experimental design course, you would learn about many variations on the analysis of variance model. These include complex experimental designs in which many independent variables are manipulated as well as designs in which repeated measures on the same individuals are taken.

CORRELATIONAL ANALYSIS AND MULTIVARIATE STATISTICS. Correlation and regression were presented in Chapter 10. Many universities offer advanced courses in correlation and regression. Chapter 10 dealt only with bivariate correlation. Bivariate correlation considers the association between two variables. Advanced courses consider the variation among three or more variables simultaneously. These are called multivariate methods. Some multivariate methods are similar to statistical tests such as the t-test. For example, discriminate analysis is analogous to performing a t-test with many dependent variables. The regression analysis described in Chapter 10 considered the prediction of a variable y from another variable x. A technique known as multiple regression considers the prediction of y from the linear combination of two or more variables. Multivariate techniques are widely used in a variety of sciences.

TEST CONSTRUCTION. Chapter 11 reviewed basic methods for evaluating reliability and validity. Some universities offer advanced courses on the construction of psychological tests. These courses focus in great detail on reliability analysis and the consequences of low reliability. In addition, they offer statistical methods for selection of items, development of fair testing procedures, and the use of tests for the selection of employees.

SURVEY RESEARCH METHODS.   Survey researchers are primarily concerned with fair and representative sampling and with evaluation of frequency data. Those interested in these problems can usually find an advanced course that deals with probability theory (as introduced in Chapter 5) and with advanced methods for the analysis of frequency data. These methods are extensions of chi-square and contingency table procedures described in Chapter 12.

These very brief sketches describe only a few of the advanced courses commonly available. If you do not go on to take advanced courses, the information presented in this text should help you understand the information in research journals. In addition you should now be able to calculate basic statistics for your own research. For those who do continue with statistical training, we hope that this book has provided some of the basic building blocks you will need for further study.

References

Ansoff, W. H. "Scales, norms, and equivalent scores." In R. L. Thorndike (ed.), *Educational Measurement,* 2nd ed. Washington, D.C.: American Council on Education, 1971.

Biderman, A. D., "The graph as a victim of adverse discrimination." *Information Decision Journal,* 1980.

Box, G. E. P., "Some theorems on quadratic forms applied in the study of analysis of variance problems. 1. Effective inequality of variance in the one-way classification." *Annals of Mathematics and Statistics,* 1954, **25**: 290–302.

Campbell, D. T., and D. W. Fiske, "Convergent and discriminant validation by the multi-trait–multimethod matrix." *Psychological Bulletin,* 1959, **56**: 81–105.

Cronbach, L. J., "Coefficient alpha and the internal structure of tests." *Psychometrika,* 1951, **16**: 297–334.

Cronbach, L. J., and P. E. Meehl, "Construct validity in psychological tests." *Psychological Bulletin,* 1955, **52**: 281–302.

Darwin, C., *The Origin of Species by Means of Natural Selection or the Preservation of Favoured Races in the Struggle for Life.* London: J. Murray, 1959.

David, F. N., *Games, Gods, and Gambling.* New York: Hafner, 1962.

Dunnette, M. D., "Aptitudes, abilities, and skills." In M. D. Dunnette, ed., *Handbook of Industrial and Organizational Psychology.* Chicago: Rand McNally, 1976.

Fisher, R. A., *Statistical Methods for Research Workers.* Edinburgh: Oliver and Boyd, 1936.

Hays, W. L., *Statistics for the Social Sciences.* New York: Holt, Rinehart and Winston, 1973.

Kaplan, R. M., J. W. Bush, and C. C. Berry, "Health status: Types of validity for an index of well-being." *Health Services Research,* 1976, **11**: 478–507.

Kaplan, R. M., and R. D. Singer, "Television violence in viewer aggression: A reevaluation of the evidence." *Journal of Social Issues,* 1976, **34**: 35–86.

Kaplan, R. M., and D. P. Saccuzzo, *Psychological Testing: Principles, Applications, and Issues.* Monterey, Calif.: Brooks/Cole, 1982.

Kirk, R. E., *Experimental Design,* 2nd ed. Monterey, Calif.: Brooks/Cole, 1982.

Kuder, G. F., and M. W. Richardson, "The theory of the estimation of reliability." *Psychometrika,* 1937, **2**: 151–160.

Landy, D., and H. Sigall, "Beauty is talent." *Journal of Personality and Social Psychology,* 1974, **29**: 299–304.

Lord, F. M., "Experimentally induced variations in Rorschach performance." *Psychological Monographs,* 1950, **64**: 10.

McCall, R., *Fundamental Statistics for Psychology,* 3rd ed. New York: Harcourt, Brace, and Jovanovich, 1980.

McNemar, Q., *Psychological Statistics,* 4th ed. New York: Wiley, 1969.

Marasculio, L. A., and M. McSweeney, *Nonparametric and Distribution-free Methods for the Social Sciences.* Monterey, Calif.: Brooks/Cole, 1977.

Messick, S., and A. Jungeblut, "Time and method in coaching for the SAT." *Psychological Bulletin,* 1981, **89**: 191–216.

Minium, E. W., and R. B. Clark, *Elements of Statistical Reasoning.* New York: Wiley, 1982.

Nunnally, J. C., *Psychometric Theory,* 2nd ed. New York: McGraw-Hill, 1978.

Pearson, K., "Mathematical contributions to the theory of evolution. III: Regression, heredity, and panmixia." *Philosophical Transactions,* 1896, **187**: 253–318.

Tukey, J. W., *Exploratory Data Analysis.* Reading, Mass.: Addison-Wesley, 1977.

Uhl, N., and T. Eisenberg, "Predicting shrinkage in the multiple correlation coefficient." *Educational and Psychological Measurement,* 1970, **30**: 487–489.

Yaremko, R. M., H. Harari, R. C. Harrison, and E. Lynn, *Reference Handbook of Research and Statistical Methods in Psychology.* New York: Harper and Row, 1982.

Wainer, H., and D. Thissen, "Graphical data analysis." *Annual Review of Psychology,* 1980, **32**: 191–241.

Appendixes

Appendix 1

Table of probabilities under the normal curve

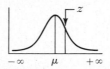

1	2	3	4	1	2	3	4
	AREA FROM	AREA FROM	AREA FROM		AREA FROM	AREA FROM	AREA FROM
Z	MEAN TO Z	−∞ TO Z	Z TO +∞	Z	MEAN TO Z	−∞ TO Z	Z TO +∞
0.00	.0000	.5000	.5000	0.25	.0987	.5987	.4013
0.01	.0040	.5040	.4960	0.26	.1026	.6026	.3974
0.02	.0080	.5080	.4920	0.27	.1064	.6064	.3936
0.03	.0120	.5120	.4880	0.28	.1103	.6103	.3897
0.04	.0160	.5160	.4840	0.29	.1141	.6141	.3859
0.05	.0199	.5199	.4801	0.30	.1179	.6179	.3821
0.06	.0239	.5239	.4761	0.31	.1217	.6217	.3783
0.07	.0279	.5279	.4721	0.32	.1255	.6255	.3745
0.08	.0319	.5319	.4681	0.33	.1293	.6293	.3707
0.09	.0359	.5359	.4641	0.34	.1331	.6331	.3669
0.10	.0398	.5398	.4602	0.35	.1368	.6368	.3632
0.11	.0438	.5438	.4562	0.36	.1406	.6406	.3594
0.12	.0478	.5478	.4522	0.37	.1443	.6443	.3557
0.13	.0517	.5517	.4483	0.38	.1480	.6480	.3520
0.14	.0557	.5557	.4443	0.39	.1517	.6517	.3483
0.15	.0596	.5596	.4404	0.40	.1554	.6554	.3446
0.16	.0636	.5636	.4364	0.41	.1591	.6591	.3409
0.17	.0675	.5675	.4325	0.42	.1628	.6628	.3372
0.18	.0714	.5714	.4286	0.43	.1664	.6664	.3336
0.19	.0753	.5753	.4247	0.44	.1700	.6700	.3300
0.20	.0793	.5793	.4207	0.45	.1736	.6736	.3264
0.21	.0832	.5832	.4168	0.46	.1772	.6772	.3228
0.22	.0871	.5871	.4129	0.47	.1808	.6808	.3192
0.23	.0910	.5910	.4090	0.48	.1844	.6844	.3156
0.24	.0948	.5948	.4052	0.49	.1879	.6879	.3121

Table of probabilities under the normal curve (*continued*)

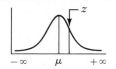

1	2	3	4	1	2	3	4
Z	AREA FROM MEAN TO Z	AREA FROM −∞ TO Z	AREA FROM Z TO +∞	Z	AREA FROM MEAN TO Z	AREA FROM −∞ TO Z	AREA FROM Z TO +∞
0.50	.1915	.6915	.3085	0.75	.2734	.7734	.2266
0.51	.1950	.6950	.3050	0.76	.2764	.7764	.2236
0.52	.1985	.6985	.3015	0.77	.2794	.7794	.2206
0.53	.2019	.7019	.2981	0.78	.2823	.7823	.2177
0.54	.2054	.7054	.2946	0.79	.2852	.7852	.2148
0.55	.2088	.7088	.2912	0.80	.2881	.7881	.2119
0.56	.2123	.7123	.2877	0.81	.2910	.7910	.2090
0.57	.2157	.7157	.2843	0.82	.2939	.7939	.2061
0.58	.2190	.7190	.2810	0.83	.2967	.7967	.2033
0.59	.2224	.7224	.2776	0.84	.2995	.7995	.2005
0.60	.2257	.7257	.2743	0.85	.3023	.8023	.1977
0.61	.2291	.7291	.2709	0.86	.3051	.8051	.1949
0.62	.2324	.7324	.2676	0.87	.3078	.8078	.1922
0.63	.2357	.7357	.2643	0.88	.3106	.8106	.1894
0.64	.2389	.7389	.2611	0.89	.3133	.8133	.1867
0.65	.2422	.7422	.2578	0.90	.3159	.8159	.1841
0.66	.2454	.7454	.2546	0.91	.3186	.8186	.1814
0.67	.2486	.7486	.2514	0.92	.3212	.8212	.1788
0.68	.2517	.7517	.2483	0.93	.3238	.8238	.1762
0.69	.2549	.7549	.2451	0.94	.3264	.8264	.1736
0.70	.2580	.7580	.2420	0.95	.3289	.8289	.1711
0.71	.2611	.7611	.2389	0.96	.3315	.8315	.1685
0.72	.2642	.7642	.2358	0.97	.3340	.8340	.1660
0.73	.2673	.7673	.2327	0.98	.3365	.8365	.1635
0.74	.2704	.7704	.2296	0.99	.3389	.8389	.1611

(*continued*)

Table of probabilities under the normal curve (*continued*)

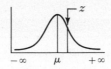

1 Z	2 AREA FROM MEAN TO Z	3 AREA FROM −∞ TO Z	4 AREA FROM Z TO +∞	1 Z	2 AREA FROM MEAN TO Z	3 AREA FROM −∞ TO Z	4 AREA FROM Z TO +∞
1.00	.3413	.8413	.1587	1.30	.4032	.9032	.0968
1.01	.3438	.8438	.1562	1.31	.4049	.9049	.0951
1.02	.3461	.8461	.1539	1.32	.4066	.9066	.0934
1.03	.3485	.8485	.1515	1.33	.4082	.9082	.0918
1.04	.3508	.8508	.1492	1.34	.4099	.9099	.0901
1.05	.3531	.8531	.1469	1.35	.4115	.9115	.0885
1.06	.3554	.8554	.1446	1.36	.4131	.9131	.0869
1.07	.3577	.8577	.1423	1.37	.4147	.9147	.0853
1.08	.3599	.8599	.1401	1.38	.4162	.9162	.0838
1.09	.3621	.8621	.1379	1.39	.4177	.9177	.0823
1.10	.3643	.8643	.1357	1.40	.4192	.9192	.0808
1.11	.3665	.8665	.1335	1.41	.4207	.9207	.0793
1.12	.3686	.8686	.1314	1.42	.4222	.9222	.0778
1.13	.3708	.8708	.1292	1.43	.4236	.9236	.0764
1.14	.3729	.8729	.1271	1.44	.4251	.9251	.0749
1.15	.3749	.8749	.1251	1.45	.4265	.9265	.0735
1.16	.3770	.8770	.1230	1.46	.4279	.9279	.0721
1.17	.3790	.8790	.1210	1.47	.4292	.9292	.0708
1.18	.3810	.8810	.1190	1.48	.4306	.9306	.0694
1.19	.3830	.8830	.1170	1.49	.4319	.9319	.0681
1.20	.3849	.8849	.1151	1.50	.4332	.9332	.0668
1.21	.3869	.8869	.1131	1.51	.4345	.9345	.0655
1.22	.3888	.8888	.1112	1.52	.4357	.9357	.0643
1.23	.3907	.8907	.1093	1.53	.4370	.9370	.0630
1.24	.3925	.8925	.1075	1.54	.4382	.9382	.0618
1.25	.3944	.8944	.1056	1.55	.4394	.9394	.0606
1.26	.3962	.8962	.1038	1.56	.4406	.9406	.0594
1.27	.3980	.8980	.1020	1.57	.4418	.9418	.0582
1.28	.3997	.8997	.1003	1.58	.4429	.9429	.0571
1.29	.4015	.9015	.0985	1.59	.4441	.9441	.0559

(*continued*)

Table of probabilities under the normal curve (*continued*)

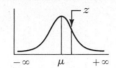

1 Z	2 AREA FROM MEAN TO Z	3 AREA FROM −∞ TO Z	4 AREA FROM Z TO +∞	1 Z	2 AREA FROM MEAN TO Z	3 AREA FROM −∞ TO Z	4 AREA FROM Z TO +∞
1.60	.4452	.9452	.0548	1.90	.4713	.9713	.0287
1.61	.4463	.9463	.0537	1.91	.4719	.9719	.0281
1.62	.4474	.9474	.0526	1.92	.4726	.9726	.0274
1.63	.4484	.9484	.0516	1.93	.4732	.9732	.0268
1.64	.4495	.9495	.0505	1.94	.4738	.9738	.0262
1.65	.4505	.9505	.0495	1.95	.4744	.9744	.0256
1.66	.4515	.9515	.0485	1.96	.4750	.9750	.0250
1.67	.4525	.9525	.0475	1.97	.4756	.9756	.0244
1.68	.4535	.9535	.0465	1.98	.4761	.9761	.0239
1.69	.4545	.9545	.0455	1.99	.4767	.9767	.0233
1.70	.4554	.9554	.0446	2.00	.4772	.9772	.0228
1.71	.4564	.9564	.0436	2.01	.4778	.9778	.0222
1.72	.4573	.9573	.0427	2.02	.4783	.9783	.0217
1.73	.4582	.9582	.0418	2.03	.4788	.9788	.0212
1.74	.4591	.9591	.0409	2.04	.4793	.9793	.0207
1.75	.4599	.9599	.0401	2.05	.4798	.9798	.0202
1.76	.4608	.9608	.0392	2.06	.4803	.9803	.0197
1.77	.4616	.9616	.0384	2.07	.4808	.9808	.0192
1.78	.4625	.9625	.0375	2.08	.4812	.9812	.0188
1.79	.4633	.9633	.0367	2.09	.4817	.9817	.0183
1.80	.4641	.9641	.0359	2.10	.4821	.9821	.0179
1.81	.4649	.9649	.0351	2.11	.4826	.9826	.0174
1.82	.4656	.9656	.0344	2.12	.4830	.9830	.0170
1.83	.4664	.9664	.0336	2.13	.4834	.9834	.0166
1.84	.4671	.9671	.0329	2.14	.4838	.9838	.0162
1.85	.4678	.9678	.0322	2.15	.4842	.9842	.0158
1.86	.4686	.9686	.0314	2.16	.4846	.9846	.0154
1.87	.4693	.9693	.0307	2.17	.4850	.9850	.0150
1.88	.4699	.9699	.0301	2.18	.4854	.9854	.0146
1.89	.4706	.9706	.0294	2.19	.4857	.9857	.0143

(*continued*)

Table of probabilities under the normal curve   (*continued*)

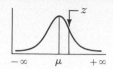

1 Z	2 AREA FROM MEAN TO Z	3 AREA FROM −∞ TO Z	4 AREA FROM Z TO +∞	1 Z	2 AREA FROM MEAN TO Z	3 AREA FROM −∞ TO Z	4 AREA FROM Z TO +∞
2.20	.4861	.9861	.0139	2.50	.4938	.9938	.0062
2.21	.4864	.9864	.0136	2.51	.4940	.9940	.0060
2.22	.4868	.9868	.0132	2.52	.4941	.9941	.0059
2.23	.4871	.9871	.0129	2.53	.4943	.9943	.0057
2.24	.4875	.9875	.0125	2.54	.4945	.9945	.0055
2.25	.4878	.9878	.0122	2.55	.4946	.9946	.0054
2.26	.4881	.9881	.0119	2.56	.4948	.9948	.0052
2.27	.4884	.9884	.0116	2.57	.4949	.9949	.0051
2.28	.4887	.9887	.0113	2.58	.4951	.9951	.0049
2.29	.4890	.9890	.0110	2.59	.4952	.9952	.0048
2.30	.4893	.9893	.0107	2.60	.4953	.9953	.0047
2.31	.4896	.9896	.0104	2.61	.4955	.9955	.0045
2.32	.4898	.9898	.0102	2.62	.4956	.9956	.0044
2.33	.4901	.9901	.0099	2.63	.4957	.9957	.0043
2.34	.4904	.9904	.0096	2.64	.4959	.9959	.0041
2.35	.4906	.9906	.0094	2.65	.4960	.9960	.0040
2.36	.4909	.9909	.0091	2.66	.4961	.9961	.0039
2.37	.4911	.9911	.0089	2.67	.4962	.9962	.0038
2.38	.4913	.9913	0087	2.68	.4963	.9963	.0037
2.39	.4916	.9916	.0084	2.69	.4964	.9964	.0036
2.40	.4918	.9918	.0082	2.70	.4965	.9965	.0035
2.41	.4920	.9920	.0080	2.71	.4966	.9966	.0034
2.42	.4922	.9922	.0078	2.72	.4967	.9967	.0033
2.43	.4925	.9925	.0075	2.73	.4968	.9968	.0032
2.44	.4927	.9927	.0073	2.74	.4969	.9969	.0031
2.45	.4929	.9929	.0071	2.75	.4970	.9970	.0030
2.46	.4931	.9931	.0069	2.76	.4971	.9971	.0029
2.47	.4932	.9932	.0068	2.77	.4972	.9972	.0028
2.48	.4934	.9934	.0066	2.78	.4973	.9973	.0027
2.49	.4936	.9936	.0064	2.79	.4974	.9974	.0026

Table of probabilities under the normal curve (*continued*)

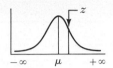

1 Z	2 AREA FROM MEAN TO Z	3 AREA FROM −∞ TO Z	4 AREA FROM Z TO +∞	1 Z	2 AREA FROM MEAN TO Z	3 AREA FROM −∞ TO Z	4 AREA FROM Z TO +∞
2.80	.4974	.9974	.0026	3.05	.4989	.9989	.0011
2.81	.4975	.9975	.0025	3.06	.4989	.9989	.0011
2.82	.4976	.9976	.0024	3.07	.4989	.9989	.0011
2.83	.4977	.9977	.0023	3.08	.4990	.9990	.0010
2.84	.4977	.9977	.0023	3.09	.4990	.9990	.0010
2.85	.4978	.9978	.0022	3.10	.4990	.9990	.0010
2.86	.4979	.9979	.0021	3.11	.4991	.9991	.0009
2.87	.4979	.9979	.0021	3.12	.4991	.9991	.0009
2.88	.4980	.9980	.0020	3.13	.4991	.9991	.0009
2.89	.4981	.9981	.0019	3.14	.4992	.9992	.0008
2.90	.4981	.9981	.0019	3.15	.4992	.9992	.0008
2.91	.4982	.9982	.0018	3.16	.4992	.9992	.0008
2.92	.4982	.9982	.0018	3.17	.4992	.9992	.0008
2.93	.4983	.9983	.0017	3.18	.4993	.9993	.0007
2.94	.4984	.9984	.0016	3.19	.4993	.9993	.0007
2.95	.4984	.9984	.0016	3.20	.4993	.9993	.0007
2.96	.4985	.9985	.0015	3.21	.4993	.9993	.0007
2.97	.4985	.9985	.0015	3.22	.4994	.9994	.0006
2.98	.4986	.9986	.0014	3.23	.4994	.9994	.0006
2.99	.4986	.9986	.0014	3.24	.4994	.9994	.0006
3.00	.4987	.9987	.0013	3.30	.4995	.9995	.0005
3.01	.4987	.9987	.0013	3.40	.4997	.9997	.0003
3.02	.4987	.9987	.0013	3.50	.4998	.9998	.0002
3.03	.4988	.9988	.0012	3.60	.4998	.9998	.0002
3.04	.4988	.9988	.0012	3.70	.4999	.9999	.0001

Adapted from A. L. Edwards, *Statistical Methods, 3rd ed.* Copyright © 1973, 1967 by Allen L. Edwards. First edition copyright 1954 by Allen L. Edwards under the title *Statistical Methods for the Behavioral Sciences.* Reprinted by permission of Holt, Rinehart and Winston.

Appendix 2

Critical values of t^a

For any given df, the table shows the values of t corresponding to various levels of probability. Obtained t is significant at a given level if it is equal to or *greater than* the value shown in the table.

df	LEVEL OF SIGNIFICANCE FOR ONE-TAILED TEST					
	.10	.05	.025	.01	.005	.0005
	LEVEL OF SIGNIFICANCE FOR TWO-TAILED TEST					
df	.20	.10	.05	.02	.01	.001
1	3.078	6.314	12.706	31.821	63.657	636.619
2	1.886	2.920	4.303	6.965	9.925	31.598
3	1.638	2.353	3.182	4.541	5.841	12.941
4	1.533	2.132	2.776	3.747	4.604	8.610
5	1.476	2.015	2.571	3.365	4.032	6.859
6	1.440	1.943	2.447	3.143	3.707	5.959
7	1.415	1.895	2.365	2.998	3.499	5.405
8	1.397	1.860	2.306	2.896	3.355	5.041
9	1.383	1.833	2.262	2.821	3.250	4.781
10	1.372	1.812	2.228	2.764	3.169	4.587
11	1.363	1.796	2.201	2.718	3.106	4.437
12	1.356	1.782	2.179	2.681	3.055	4.318
13	1.350	1.771	2.160	2.650	3.012	4.221
14	1.345	1.761	2.145	2.624	2.977	4.140
15	1.341	1.753	2.131	2.602	2.947	4.073
16	1.337	1.746	2.120	2.583	2.921	4.015
17	1.333	1.740	2.110	2.567	2.898	3.965
18	1.330	1.734	2.101	2.552	2.878	3.922
19	1.328	1.729	2.093	2.539	2.861	3.883
20	1.325	1.725	2.086	2.528	2.845	3.850
21	1.323	1.721	2.080	2.518	2.831	3.819
22	1.321	1.717	2.074	2.508	2.819	3.792
23	1.319	1.714	2.069	2.500	2.807	3.767
24	1.318	1.711	2.064	2.492	2.797	3.745
25	1.316	1.708	2.060	2.485	2.787	3.725

(continued)

Critical values of t [a]   (*continued*)

df	LEVEL OF SIGNIFICANCE FOR ONE-TAILED TEST					
	.10	.05	.025	.01	.005	.0005
	LEVEL OF SIGNIFICANCE FOR TWO-TAILED TEST					
	.20	.10	.05	.02	.01	.001
26	1.315	1.706	2.056	2.479	2.779	3.707
27	1.314	1.703	2.052	2.473	2.771	3.690
28	1.313	1.701	2.048	2.467	2.763	3.674
29	1.311	1.699	2.045	2.462	2.756	3.659
30	1.310	1.697	2.042	2.457	2.750	3.646
40	1.303	1.684	2.021	2.423	2.704	3.551
60	1.296	1.671	2.000	2.390	2.660	3.460
120	1.289	1.658	1.980	2.358	2.617	3.373
∞	1.282	1.645	1.960	2.326	2.576	3.291

[a]Appendix 2 is taken from Table III of Fisher and Yates, *Statistical tables for biological, agricultural and medical research,* published by Longman Group Ltd., London (previously published by Oliver and Boyd, Edinburgh), and by permission of the authors and publishers.

Appendix 3

Critical values of F ($\alpha = .05$ first row; $\alpha = .01$ second row)											

n_2 df FOR DE-NOMI-NATOR	n_1 df FOR NUMERATOR											
	1	2	3	4	5	6	7	8	9	10	11	12
1	161	200	216	225	230	234	237	239	241	242	243	244
	4,052	4,999	5,403	5,625	5,764	5,859	5,928	5,981	6,022	6,056	6,082	6,106
2	18.51	19.00	19.16	19.25	19.30	19.33	19.36	19.37	19.38	19.39	19.40	19.41
	98.49	99.00	99.17	99.25	99.30	99.33	99.34	99.36	99.38	99.40	99.41	99.42
3	10.13	9.55	9.28	9.12	9.01	8.94	8.88	8.84	8.81	8.78	8.76	8.74
	34.12	30.82	29.46	28.71	28.24	27.91	27.67	27.49	27.34	27.23	27.13	27.05
4	7.71	6.94	6.59	6.39	6.26	6.16	6.09	6.04	6.00	5.96	5.93	5.91
	21.20	18.00	16.69	15.98	15.52	15.21	14.98	14.80	14.66	14.54	14.45	14.37
5	6.61	5.79	5.41	5.19	5.05	4.95	4.88	4.82	4.78	4.74	4.70	4.68
	16.26	13.27	12.06	11.39	10.97	10.67	10.45	10.27	10.15	10.05	9.96	9.89
6	5.99	5.14	4.76	4.53	4.39	4.28	4.21	4.15	4.10	4.06	4.03	4.00
	13.74	10.92	9.78	9.15	8.75	8.47	8.26	8.10	7.98	7.87	7.79	7.72
7	5.59	4.74	4.35	4.12	3.97	3.87	3.79	3.73	3.68	3.63	3.60	3.57
	12.25	9.55	8.45	7.85	7.46	7.19	7.00	6.84	6.71	6.62	6.54	6.47
8	5.32	4.46	4.07	3.84	3.69	3.58	3.50	3.44	3.39	3.34	3.31	3.28
	11.26	8.65	7.59	7.01	6.63	6.37	6.19	6.03	5.91	5.82	5.74	5.67
9	5.12	4.26	3.86	3.63	3.48	3.37	3.29	3.23	3.18	3.13	3.10	3.07
	10.56	8.02	6.99	6.42	6.06	5.80	5.62	5.47	5.35	5.26	5.18	5.11
10	4.96	4.10	3.71	3.48	3.33	3.22	3.14	3.07	3.02	2.97	2.94	2.91
	10.04	7.56	6.55	5.99	5.64	5.39	5.21	5.06	4.95	4.85	4.78	4.71
11	4.84	3.98	3.59	3.36	3.20	3.09	3.01	2.95	2.90	2.86	2.82	2.79
	9.65	7.20	6.22	5.67	5.32	5.97	4.88	4.74	4.63	4.54	4.46	4.40
12	4.75	3.88	3.49	3.26	3.11	3.00	2.92	2.85	2.80	2.76	2.72	2.69
	9.33	6.93	5.95	5.41	5.06	4.82	4.65	4.50	4.39	4.30	4.22	4.16
13	4.67	3.80	3.41	3.18	3.02	2.92	2.84	2.77	2.72	2.67	2.63	2.60
	9.07	6.70	5.74	5.20	4.86	4.62	4.44	4.30	4.19	4.10	4.02	3.96

(continued)

| Critical values of F ($\alpha = .05$ first row; $\alpha = .01$ second row) *(continued)* |

n_1 *df* FOR NUMERATOR

14	16	20	24	30	40	50	75	100	200	500	∞
245	246	248	249	250	251	252	253	253	254	254	254
6,142	6,169	6,208	6,234	6,258	6,286	6,302	6,323	6,334	6,352	6,361	6,366
19.42	19.43	19.44	19.45	19.46	19.47	19.47	19.48	19.49	19.49	19.50	19.50
99.43	99.44	99.45	99.46	99.47	99.48	99.48	99.49	99.49	99.40	99.50	99.50
8.71	8.69	8.66	8.64	8.62	8.60	8.58	8.57	8.56	8.54	8.54	8.53
26.92	26.83	26.69	26.60	26.50	26.41	26.35	26.27	26.23	26.18	26.14	26.12
5.87	5.84	5.80	5.77	5.74	5.71	5.70	5.68	5.66	5.65	5.64	5.63
14.24	14.15	14.02	13.93	13.83	13.74	13.69	13.61	13.57	13.52	13.48	13.46
4.64	4.60	5.46	4.53	4.50	4.46	4.44	4.42	4.40	4.38	4.37	4.36
9.77	9.68	9.55	9.47	9.38	9.29	9.24	9.17	9.13	9.07	9.04	9.02
3.96	3.92	3.87	3.84	3.81	3.77	3.75	3.72	3.71	3.69	3.68	3.67
7.60	7.52	7.39	7.31	7.23	7.14	7.09	7.02	6.99	6.94	6.90	6.88
3.52	3.49	3.44	3.41	3.38	3.34	3.32	3.29	3.28	3.25	3.24	3.23
6.35	6.27	6.15	6.07	5.98	5.90	5.85	5.78	5.75	5.70	5.67	5.65
3.23	3.20	3.15	3.12	3.08	3.05	3.03	3.00	2.98	2.96	2.94	2.93
5.56	5.48	5.36	5.28	5.20	5.11	5.06	5.00	4.96	4.91	4.88	4.86
3.02	2.98	2.93	2.90	2.86	2.82	2.80	2.77	2.76	2.73	2.72	2.71
5.00	4.92	4.80	4.73	4.64	4.56	4.51	4.45	4.41	4.36	4.33	4.31
2.86	2.82	2.77	2.74	2.70	2.67	2.64	2.61	2.59	2.56	2.55	2.54
4.60	4.52	4.41	4.33	4.25	4.17	4.12	4.05	4.01	3.96	3.93	3.91
2.74	2.70	2.65	2.61	2.57	2.53	2.50	2.47	2.45	2.42	2.41	2.40
4.29	4.21	4.10	4.02	3.94	3.86	3.80	3.74	3.70	3.66	3.62	3.60
2.64	2.60	2.54	2.50	2.46	2.42	2.40	2.36	2.35	2.32	2.31	2.30
4.05	3.98	3.86	3.78	3.70	3.61	3.56	3.49	3.46	3.41	3.38	3.36
2.55	2.51	2.46	2.42	2.38	2.34	2.32	2.28	2.26	2.24	2.22	2.21
3.85	3.78	3.67	3.59	3.51	3.42	3.37	3.30	3.27	3.21	3.18	3.16

(continued)

Critical values of F ($\alpha = .05$ first row; $\alpha = .01$ second row) (*continued*)

n_2 df FOR DE-NOMI-NATOR	n_1 df FOR NUMERATOR											
	1	2	3	4	5	6	7	8	9	10	11	12
14	4.60	3.74	3.34	3.11	2.96	2.85	2.77	2.70	2.65	2.60	2.56	2.53
	8.86	6.51	5.56	5.03	4.69	4.46	4.28	4.14	4.03	3.94	3.86	3.80
15	4.54	3.68	3.29	3.06	2.90	2.79	2.70	2.64	2.59	2.55	2.51	2.48
	8.68	6.36	5.42	4.89	4.56	4.32	4.14	4.00	3.89	3.80	3.73	3.67
16	4.49	3.63	3.24	3.01	2.85	2.74	2.66	2.59	2.54	2.49	2.45	2.42
	8.53	6.23	5.29	4.77	4.44	4.20	4.03	3.89	3.78	3.69	3.61	3.55
17	4.45	3.59	3.20	2.96	2.81	2.70	2.62	2.55	2.50	2.45	2.41	2.38
	8.40	6.11	5.18	4.67	4.34	4.10	3.93	3.79	3.68	3.59	3.52	3.45
18	4.41	3.55	3.16	2.93	2.77	2.66	2.58	2.51	2.46	2.41	2.37	2.34
	8.28	6.01	5.09	4.58	4.25	4.01	3.85	3.71	3.60	3.51	3.44	3.37
19	4.38	3.52	3.13	2.90	2.74	2.63	2.55	2.48	2.43	2.38	2.34	2.31
	8.18	5.93	5.01	4.50	4.17	3.94	3.77	3.63	3.52	3.43	3.36	3.30
20	4.35	3.49	3.10	2.87	2.71	2.60	2.52	2.45	2.40	2.35	2.31	2.28
	8.10	5.85	4.94	4.43	4.10	3.87	3.71	3.56	3.45	3.37	3.30	3.23
21	4.32	3.47	3.07	2.84	2.68	2.57	2.49	2.42	2.37	2.32	2.28	2.25
	8.02	5.78	4.87	4.37	4.04	3.81	3.65	3.51	3.40	3.31	3.24	3.17
22	4.30	3.44	3.05	2.82	2.66	2.55	2.47	2.40	2.35	2.30	2.26	2.23
	7.94	5.72	4.82	4.31	3.99	3.76	3.59	3.45	3.35	3.26	3.18	3.12
23	4.28	3.42	3.03	2.80	2.64	2.53	2.45	2.38	2.32	2.28	2.24	2.20
	7.88	5.66	4.76	4.26	3.94	3.71	3.54	3.41	3.30	3.21	3.14	3.07
24	4.26	3.40	3.01	2.78	2.62	2.51	2.43	2.36	2.30	2.26	2.22	2.18
	7.82	5.61	4.72	4.22	3.90	3.67	3.50	3.36	3.25	3.17	3.09	3.03
25	4.24	3.38	2.99	2.76	2.60	2.49	2.41	2.34	2.28	2.24	2.20	2.16
	7.77	5.57	4.68	4.18	3.86	3.63	3.46	3.32	3.21	3.13	3.05	2.99
26	4.22	3.37	2.98	2.74	2.59	2.47	2.39	2.32	2.27	2.22	2.18	2.15
	7.72	5.53	4.64	4.14	3.82	3.59	3.42	3.29	3.17	3.09	3.02	2.96

(*continued*)

Critical values of F (α = .05 first row; α = .01 second row) (*continued*)

	n_1 *df* FOR NUMERATOR										
14	16	20	24	30	40	50	75	100	200	500	∞
2.48	2.44	2.39	2.35	2.31	2.27	2.24	2.21	2.19	2.16	2.14	2.13
3.70	3.62	3.51	3.43	3.34	3.26	3.21	3.14	3.11	3.06	3.02	3.00
2.43	2.39	2.33	2.29	2.25	2.21	2.18	2.15	2.12	2.10	2.08	2.07
3.56	3.48	3.36	3.29	3.20	3.12	3.07	3.00	2.97	2.92	2.89	2.87
2.37	2.33	2.28	2.24	2.20	2.16	2.13	2.09	2.07	2.04	2.02	2.01
3.45	3.37	3.25	3.18	3.10	3.01	2.96	2.89	2.86	2.80	2.77	2.75
2.33	2.29	2.23	2.19	2.15	2.11	2.08	2.04	2.02	1.99	1.97	1.96
3.35	3.27	3.16	3.08	3.00	2.92	2.86	2.79	2.76	2.70	2.67	2.65
2.29	2.25	2.19	2.15	2.11	2.07	2.04	2.00	1.98	1.95	1.93	1.92
3.27	3.19	3.07	3.00	2.91	2.81	2.78	2.71	2.68	2.62	2.59	2.57
2.26	2.21	2.15	2.11	2.07	2.02	2.00	1.96	1.94	1.91	1.90	1.88
3.19	3.12	3.00	2.92	2.84	2.76	2.70	2.63	2.60	2.54	2.51	2.49
2.23	2.18	2.12	2.08	2.04	1.99	1.96	1.92	1.90	1.87	1.85	1.84
3.13	3.05	2.94	2.86	2.77	2.69	2.63	2.56	2.53	2.47	2.44	2.42
2.20	2.15	2.09	2.05	2.00	1.96	1.93	1.89	1.87	1.84	1.82	1.81
3.07	2.99	2.88	2.80	2.72	2.63	2.58	2.51	2.47	2.42	2.38	2.36
2.18	2.13	2.07	2.03	1.98	1.93	1.91	1.87	1.84	1.81	1.80	1.78
3.02	2.94	2.83	2.75	2.67	2.58	2.53	2.46	2.42	2.37	2.33	2.31
2.14	2.10	2.04	2.00	1.96	1.91	1.88	1.84	1.82	1.79	1.77	1.76
2.97	2.89	2.78	2.70	2.62	2.53	2.48	2.41	2.37	2.32	2.28	2.26
2.13	2.09	2.02	1.98	1.94	1.89	1.86	1.82	1.80	1.76	1.74	1.73
2.93	2.85	2.74	2.66	2.58	2.49	2.44	2.36	2.33	2.27	2.23	2.21
2.11	2.06	2.00	1.96	1.92	1.87	1.84	1.80	1.77	1.74	1.72	1.71
2.89	2.81	2.70	2.62	2.54	2.45	2.40	2.32	2.29	2.23	2.19	2.17
2.10	2.05	1.99	1.95	1.90	1.85	1.82	1.78	1.76	1.72	1.70	1.69
2.86	2.77	2.66	2.58	2.50	2.41	2.36	2.28	2.25	2.19	2.15	2.13

(continued)

Critical values of F ($\alpha = .05$ first row; $\alpha = .01$ second row)   (*continued*)

n_2 df FOR DENOMINATOR	\n_1 df FOR NUMERATOR											
	1	2	3	4	5	6	7	8	9	10	11	12
27	4.21	3.35	2.96	2.73	2.57	2.46	2.37	2.30	2.25	2.20	2.16	2.13
	7.68	5.49	4.60	4.11	3.79	3.56	3.39	3.26	3.14	3.06	2.98	2.93
28	4.20	3.34	2.95	2.71	2.56	2.44	2.36	2.29	2.24	2.19	2.15	2.12
	7.64	5.45	4.57	4.07	3.76	3.53	3.36	3.23	3.11	3.03	2.95	2.90
29	4.18	3.33	2.93	2.70	2.54	2.43	2.35	2.28	2.22	2.18	2.14	2.10
	7.60	5.42	4.54	4.04	3.73	3.50	3.33	3.20	3.08	3.00	2.92	2.87
30	4.17	3.32	2.92	2.69	2.53	2.42	2.34	2.27	2.21	2.16	2.12	2.09
	7.56	5.39	4.51	4.02	3.70	3.47	3.30	3.17	3.06	2.98	2.90	2.84
32	4.15	3.30	2.90	2.67	2.51	2.40	2.32	2.25	2.19	2.14	2.10	2.07
	7.50	5.34	4.46	3.97	3.66	3.42	3.25	3.12	3.01	2.94	2.86	2.80
34	4.13	3.28	2.88	2.65	2.49	2.38	2.30	2.23	2.17	2.12	2.08	2.05
	7.44	5.29	4.42	3.93	3.61	3.38	3.21	3.08	2.97	2.89	2.82	2.76
36	4.11	3.26	2.86	2.63	2.48	2.36	2.28	2.21	2.15	2.10	2.06	2.03
	7.39	5.25	4.38	3.89	3.58	3.35	3.18	3.04	2.94	2.86	2.78	2.72
38	4.10	3.25	2.85	2.62	2.46	2.35	2.26	2.19	2.14	2.09	2.05	2.02
	7.35	5.21	4.34	3.86	3.54	3.32	3.15	3.02	2.91	2.82	2.75	2.69
40	4.08	3.23	2.84	2.61	2.45	2.34	2.25	2.18	2.12	2.07	2.04	2.00
	7.31	5.18	4.31	3.83	3.51	3.29	3.12	2.99	2.88	2.80	2.73	2.66
42	4.07	3.22	2.83	2.59	2.44	2.32	2.24	2.17	2.11	2.06	2.02	1.99
	7.27	5.15	4.29	3.80	3.49	3.26	3.10	2.96	2.86	2.77	2.70	2.64
44	4.06	3.21	2.82	2.58	2.43	2.31	2.23	2.16	2.10	2.05	2.01	1.98
	7.24	5.12	4.26	3.78	3.46	3.24	3.07	2.94	2.84	2.75	2.68	2.62
46	4.05	3.20	2.81	2.57	2.42	2.30	2.22	2.14	2.09	2.04	2.00	1.97
	7.21	5.10	4.24	3.76	3.44	3.22	3.05	2.92	2.82	2.73	2.66	2.60
48	4.04	3.19	2.80	2.56	2.41	2.30	2.21	2.14	2.08	2.03	1.99	1.96
	7.19	5.08	4.22	3.74	3.42	3.20	3.04	2.90	2.80	2.71	2.64	2.58

(continued)

Critical values of F (α = .05 first row; α = .01 second row)   (*continued*)

	n_1 *df* FOR NUMERATOR											
	14	16	20	24	30	40	50	75	100	200	500	∞
	2.08	2.03	1.97	1.93	1.88	1.84	1.80	1.76	1.74	1.71	1.68	1.67
	2.83	2.74	2.63	2.55	2.47	2.38	2.25	2.21	2.16	2.16	2.12	2.10
	2.05	2.02	1.96	1.91	1.87	1.81	1.78	1.75	1.72	1.69	1.67	1.65
	2.80	2.71	2.60	2.52	2.44	2.35	2.30	2.22	2.18	2.13	2.09	2.06
	2.05	2.00	1.94	1.90	1.85	1.80	1.77	1.73	1.71	1.68	1.65	1.64
	2.77	2.68	2.57	2.49	2.41	2.32	2.27	2.19	2.15	2.10	2.06	2.03
	2.04	1.99	1.93	1.89	1.84	1.79	1.76	1.72	1.69	1.66	1.64	1.62
	2.74	2.66	2.55	2.47	2.38	2.29	2.24	2.16	2.13	2.07	2.03	2.01
	2.02	1.97	1.91	1.86	1.82	1.76	1.74	1.69	1.67	1.64	1.61	1.59
	2.70	2.62	2.51	2.42	2.34	2.25	2.20	2.12	2.08	2.02	1.98	1.96
	2.00	1.95	1.89	1.84	1.80	1.74	1.71	1.67	1.64	1.61	1.59	1.57
	2.66	2.58	2.47	2.38	2.30	2.21	2.15	2.08	2.04	1.98	1.94	1.91
	1.98	1.93	1.87	1.82	1.78	1.72	1.69	1.65	1.62	1.59	1.56	1.55
	2.62	2.54	2.43	2.35	2.26	2.17	2.12	2.04	2.00	1.94	1.90	1.87
	1.96	1.92	1.85	1.80	1.76	1.71	1.67	1.63	1.60	1.57	1.54	1.53
	2.59	2.51	2.40	2.32	2.22	2.14	2.08	2.00	1.97	1.90	1.86	1.84
	1.95	1.90	1.84	1.79	1.74	1.69	1.66	1.61	1.59	1.55	1.53	1.51
	2.56	2.49	2.37	2.29	2.20	2.11	2.05	1.97	1.94	1.88	1.84	1.81
	1.94	1.89	1.82	1.78	1.73	1.68	1.64	1.60	1.57	1.54	1.51	1.49
	2.54	2.46	2.35	2.26	2.17	2.08	2.02	1.94	1.91	1.85	1.80	1.78
	1.92	1.88	1.81	1.76	1.72	1.66	1.63	1.58	1.56	1.52	1.50	1.48
	2.52	2.44	2.32	2.24	2.15	2.06	2.00	1.92	1.88	1.81	1.78	1.75
	1.91	1.87	1.80	1.75	1.71	1.65	1.62	1.57	1.54	1.51	1.48	1.46
	2.50	2.42	2.30	2.22	2.13	2.04	1.98	1.90	1.86	1.80	1.76	1.72
	1.90	1.86	1.79	1.74	1.70	1.64	1.61	1.56	1.53	1.50	1.47	1.45
	2.43	2.40	2.28	2.20	2.11	2.02	1.96	1.88	1.84	1.78	1.73	1.70

(*continued*)

Critical values of *F* (α = .05 first row; α = .01 second row)   (*continued*)

n_2 df FOR DE-NOMI-NATOR	n_1 df FOR NUMERATOR 1	2	3	4	5	6	7	8	9	10	11	12
50	4.03	3.18	2.79	2.56	2.40	2.29	2.20	2.13	2.07	2.02	1.98	1.95
	7.17	5.06	4.20	3.72	3.41	3.18	3.02	2.88	2.78	2.70	2.62	2.56
55	4.02	3.17	2.78	2.54	2.38	2.27	2.18	2.11	2.05	2.00	1.97	1.93
	7.12	5.01	4.16	3.68	3.37	3.15	2.98	2.85	2.75	2.66	2.59	2.53
60	4.00	3.15	2.76	2.52	2.37	2.25	2.17	2.10	2.04	1.99	1.95	1.92
	7.08	4.98	4.13	3.65	3.34	3.12	2.95	2.82	2.72	2.63	2.56	2.50
65	3.99	3.14	2.75	2.51	2.36	2.24	2.15	2.08	2.02	1.98	1.94	1.90
	7.04	4.95	4.10	3.62	3.31	3.09	2.93	2.79	2.70	2.61	2.54	2.47
70	3.98	3.13	2.74	2.50	2.35	2.23	2.14	2.07	2.01	1.97	1.93	1.89
	7.01	4.92	4.08	3.60	3.29	3.07	2.91	2.77	2.67	2.59	2.51	2.45
80	3.96	3.11	2.72	2.48	2.33	2.21	2.12	2.05	1.99	1.95	1.91	1.88
	6.96	4.88	4.04	3.56	3.25	3.04	2.87	2.74	2.64	2.55	2.48	2.41
100	3.94	3.09	2.70	2.46	2.30	2.19	2.10	2.03	1.97	1.92	1.88	1.85
	6.90	4.82	3.98	3.51	3.20	2.99	2.82	2.69	2.59	2.51	2.43	2.36
125	3.92	3.07	2.68	2.44	2.29	2.17	2.08	2.01	1.95	1.90	1.86	1.83
	6.84	4.78	3.94	3.47	3.17	2.95	2.79	2.65	2.56	2.47	2.40	2.33
150	3.91	3.06	2.67	2.43	2.27	2.16	2.07	2.00	1.94	1.89	1.85	1.82
	6.81	4.75	3.91	3.44	3.14	2.92	2.76	2.62	2.53	2.44	2.37	2.30
200	3.89	3.04	2.65	2.41	2.26	2.14	2.05	1.98	1.92	1.87	1.83	1.80
	6.76	4.71	3.88	3.41	3.11	2.90	2.73	2.60	2.50	2.41	2.34	2.28
400	3.86	3.02	2.62	2.39	2.23	2.12	2.03	1.96	1.90	1.85	1.81	1.78
	6.70	4.66	3.83	3.36	3.06	2.85	2.69	2.55	2.46	2.37	2.29	2.23
1000	3.85	3.00	2.61	2.38	2.22	2.10	2.02	1.95	1.89	1.84	1.80	1.76
	6.66	4.62	3.80	3.34	3.04	2.82	2.66	2.53	2.43	2.34	2.26	2.20
∞	3.84	2.99	2.60	2.37	2.21	2.09	2.01	1.94	1.88	1.83	1.79	1.75
	6.64	4.60	3.78	3.32	3.02	2.80	2.64	2.51	2.41	2.32	2.24	2.18

(*continued*)

Critical values of F ($\alpha = .05$ first row; $\alpha = .01$ second row)   (*continued*)

						n_1 df FOR NUMERATOR						
14	16	20	24	30	40	50	75	100	200	500	∞
1.90	1.85	1.78	1.74	1.69	1.63	1.60	1.55	1.52	1.48	1.46	1.44
2.46	2.39	2.26	2.18	2.10	2.00	1.94	1.86	1.82	1.76	1.71	1.68
1.88	1.83	1.76	1.72	1.67	1.61	1.58	1.52	1.50	1.46	1.43	1.41
2.43	2.35	2.23	2.15	2.06	1.96	1.90	1.82	1.78	1.71	1.66	1.64
1.86	1.81	1.75	1.70	1.65	1.59	1.56	1.50	1.48	1.44	1.41	1.39
2.40	2.32	2.20	2.12	2.03	1.93	1.87	1.79	1.74	1.68	1.63	1.60
1.85	1.80	1.73	1.68	1.63	1.57	1.54	1.49	1.46	1.42	1.39	1.37
2.37	2.30	2.18	2.09	2.00	1.90	1.84	1.76	1.71	1.64	1.60	1.56
1.84	1.79	1.72	1.67	1.62	1.56	1.53	1.47	1.45	1.40	1.37	1.35
2.35	2.28	2.15	2.07	1.98	1.88	1.82	1.74	1.69	1.62	1.56	1.53
1.82	1.77	1.70	1.65	1.60	1.54	1.51	1.45	1.42	1.38	1.35	1.32
2.32	2.24	2.11	2.03	1.94	1.84	1.78	1.70	1.65	1.57	1.52	1.49
1.79	1.75	1.68	1.63	1.57	1.51	1.48	1.42	1.39	1.34	1.30	1.28
2.26	2.19	2.06	1.98	1.89	1.79	1.73	1.64	1.59	1.51	1.46	1.43
1.77	1.72	1.65	1.60	1.55	1.49	1.45	1.39	1.36	1.31	1.27	1.25
2.23	2.15	2.03	1.94	1.85	1.75	1.68	1.59	1.54	1.46	1.40	1.37
1.76	1.71	1.64	1.59	1.54	1.47	1.44	1.37	1.34	1.29	1.25	1.22
2.20	2.12	2.00	1.91	1.83	1.72	1.66	1.56	1.51	1.43	1.37	1.33
1.74	1.69	1.62	1.57	1.52	1.45	1.42	1.35	1.32	1.26	1.22	1.19
2.17	2.09	1.97	1.88	1.79	1.69	1.62	1.53	1.48	1.39	1.33	1.28
1.72	1.67	1.60	1.54	1.49	1.42	1.38	1.32	1.28	1.22	1.16	1.13
2.12	2.04	1.92	1.84	1.74	1.64	1.57	1.47	1.42	1.32	1.24	1.19
1.70	1.65	1.58	1.53	1.47	1.41	1.36	1.30	1.26	1.19	1.13	1.08
2.09	2.01	1.89	1.81	1.71	1.61	1.54	1.44	1.38	1.28	1.19	1.11
1.69	1.64	1.57	1.52	1.46	1.40	1.35	1.28	1.24	1.17	1.11	1.00
2.07	1.99	1.87	1.79	1.69	1.59	1.52	1.41	1.36	1.25	1.15	1.00

Appendix 4

Critical values of r for $\alpha = .05$ and $\alpha = .01$ *(two-tailed test)*.		
df^*	$\alpha = .05$	$\alpha = .01$
1	.99692	.999877
2	.9500	.99000
3	.878	.9587
4	.811	.9172
5	.754	.875
6	.707	.834
7	.666	.798
8	.632	.765
9	.602	.735
10	.576	.708
11	.553	.684
12	.532	.661
13	.514	.641
14	.497	.623
15	.482	.606
16	.468	.590
17	.456	.575
18	.444	.561
19	.433	.549
20	.423	.537
25	.381	.487
30	.349	.449
35	.325	.418
40	.304	.393
45	.288	.372
50	.273	.354
60	.250	.325
70	.232	.302
80	.217	.283
90	.205	.267
100	.195	.254

*df are equal to $N - 2$ where N is the number of paired observations.

Reprinted with permission from Table IX.1, Percentage Points. Distribution of the Correlation Coefficient. When $\rho = 0$, *CRC Handbook of Tables for Probability and Statistics* (2nd ed.). Copyright 1968, CRC Press, Inc., Boca Raton, Florida.

Critical values of the chi-square distribution for $\alpha = .05$ and $\alpha = .01$.		
df	$\alpha = .05$	$\alpha = .01$
1	3.84	6.63
2	5.99	9.21
3	7.81	11.3
4	9.49	13.3
5	11.1	15.1
6	12.6	16.8
7	14.1	18.5
8	15.5	20.1
9	16.9	21.7
10	18.3	23.2
11	19.7	24.7
12	21.0	26.2
13	22.4	27.7
14	23.7	29.1
15	25.0	30.6
16	26.3	32.0
17	27.6	33.4
18	28.9	34.8
19	30.1	36.2
20	31.4	37.6
21	32.7	38.9
22	33.9	40.3
23	35.2	41.6
24	36.4	43.0
25	37.7	44.3
26	38.9	45.6
27	40.1	47.0
28	41.3	48.3
29	42.6	49.6
30	43.8	50.9

Reprinted with permission from Table V.1, Percentage Points, Chi-Square Distribution. *CRC Handbook of Tables for Probability and Statistics,* (2nd ed.). Copyright 1968, CRC Press, Inc., Boca Raton, Florida.

Appendix 6

Critical Values of ΣR_X for the Mann-Whitney Test[1]

n_1	.005	.01	.025	.05	n_1	.005	.01	.025	.05
		$n_2 = 3$					$n_2 = 4$		
3				6–15	4			10–26	11–25
4				6–18	5		10–30	11–29	12–28
5			6–21	7–20	6	10–34	11–33	12–32	13–31
6			7–23	8–22	7	10–38	11–37	13–35	14–34
7		6–27	7–26	8–25	8	11–41	12–40	14–48	15–37
8		6–30	8–28	9–27	9	11–45	13–43	14–42	16–40
9	6–33	7–32	8–31	10–29	10	12–48	13–47	15–45	17–43
10	6–36	7–35	9–33	10–32	11	12–52	14–50	16–48	18–46
11	6–39	7–38	9–36	11–34	12	13–55	15–53	17–51	19–49
12	7–41	8–40	10–38	11–37	13	13–59	15–57	18–54	20–52
13	7–44	8–43	10–41	12–39	14	14–62	16–60	19–57	21–55
14	7–47	8–46	11–43	13–41	15	15–65	17–63	20–60	22–58
15	8–49	9–48	11–46	13–44					
		$n_2 = 5$					$n_2 = 6$		
5	15–40	16–39	17–38	19–36	6	23–55	24–54	26–52	28–50
6	16–44	17–43	18–42	20–40	7	24–60	25–59	27–57	29–55
7	16–49	18–47	20–45	21–44	8	25–65	27–63	29–61	31–59
8	17–53	19–51	21–49	23–47	9	26–70	28–68	31–65	33–63
9	18–57	20–55	22–53	24–51	10	27–75	29–73	32–70	35–67
10	19–61	21–59	23–57	26–54	11	28–80	30–78	34–74	37–71
11	20–65	22–63	24–61	27–58	12	30–84	32–82	35–79	38–76
12	21–69	23–67	26–64	28–62	13	31–89	33–87	37–83	40–80
13	22–73	24–71	27–68	30–65	14	32–94	34–92	38–88	42–84
14	22–78	25–75	28–72	31–69	15	33–99	36–96	40–92	44–88
15	23–82	26–79	29–76	33–72					

[1]Adaptation of Table 1 in L. R. Verdooren, "Extended Tables of Critical Values for Wilcoxon's Test Statistic," *Biometrika,* **50,** 177–186 (1963), by permission of the author and the Biometrika Trustees.

Critical Values of ΣR_X for the Mann-Whitney Test[1]   (*continued*)

	$n_2 = 7$					$n_2 = 8$			
n_1	.005	.01	.025	.05	n_1	.005	.01	.025	.05
7	32–73	34–71	36–69	39–66	8	43–93	45–91	49–87	51–85
8	34–78	35–77	38–74	41–71	9	45–99	47–97	51–93	54–90
9	35–84	37–82	40–79	43–76	10	47–105	49–103	53–99	56–96
10	37–89	39–87	42–84	45–81	11	49–111	51–109	55–105	59–101
11	38–95	40–93	44–89	47–86	12	51–117	53–115	58–110	62–106
12	40–100	42–98	46–94	49–91	13	53–123	56–120	60–116	64–112
13	41–106	44–103	48–99	52–95	14	54–130	58–126	62–122	67–117
14	43–111	45–109	50–104	54–100	15	56–136	60–132	65–127	69–123
15	44–117	47–114	52–109	56–105					

	$n_2 = 9$					$n_2 = 10$			
n_1	.005	.01	.025	.05	n_1	.005	.01	.025	.05
9	56–115	59–112	62–109	66–105	10	71–139	74–136	78–132	82–128
10	58–122	61–119	65–115	69–111	11	73–147	77–143	81–139	86–134
11	61–128	63–126	68–121	72–117	12	76–154	79–151	84–146	89–141
12	63–135	66–132	71–127	75–123	13	79–161	82–158	88–152	92–148
13	65–142	68–139	73–134	78–129	14	81–169	85–165	91–159	96–154
14	67–149	71–145	76–140	81–135	15	84–176	88–172	94–166	99–161
15	69–156	73–152	79–146	84–141					

	$n_2 = 11$					$n_2 = 12$			
n_1	.005	.01	.025	.05	n_1	.005	.01	.025	.05
11	87–166	91–162	96–157	100–153	12	105–195	109–191	115–185	120–180
12	90–174	94–170	99–165	104–160	13	109–203	113–199	119–193	125–187
13	93–182	97–178	103–172	108–167	14	112–212	116–208	123–201	129–195
14	96–190	100–186	106–180	112–174	15	115–221	120–216	127–209	133–203
15	99–198	103–194	110–187	116–181					

	$n_2 = 13$					$n_2 = 14$			
n_1	.005	.01	.025	.05	n_1	.005	.01	.025	.05
13	125–226	130–221	136–215	142–209	14	147–259	152–254	160–246	166–240
14	129–235	134–230	141–223	147–217	15	151–269	156–264	164–256	171–249
15	133–244	138–239	145–232	152–225					

	$n_2 = 15$			
n_1	.005	.01	.025	.05
15	171–294	176–289	184–281	192–273

Appendix 7

Percentage points for the Kruskal–Wallis Test for $K = 3$ and $n \leq 5$

SAMPLE SIZES					
n_1	n_2	n_3	$\alpha \leq .10$	$\alpha \leq .05$	$\alpha \leq .01$
2	2	2	4.57	—	—
3	2	1	4.29	—	—
3	2	2	—	4.71	—
3	3	1	4.57	5.14	—
3	3	2	4.56	5.36	—
3	3	3	4.62	5.60	7.20
4	2	1	4.50	—	—
4	2	2	4.46	5.33	—
4	3	1	4.06	5.21	—
4	3	2	4.51	5.44	6.44
4	3	3	4.71	5.73	6.75
4	4	1	4.17	4.97	6.67
4	4	2	4.55	5.45	7.04
4	4	3	4.55	5.60	7.14
4	4	4	4.65	5.69	7.66
5	2	1	4.20	5.00	—
5	2	2	4.36	5.16	6.53
5	3	1	4.02	4.96	—
5	3	2	4.65	5.25	6.82
5	3	3	4.53	5.65	7.08
5	4	1	3.99	4.99	6.95
5	4	2	4.54	5.27	7.12
5	4	3	4.55	5.63	7.44
5	4	4	4.62	5.62	7.76
5	5	1	4.11	5.13	7.31
5	5	2	4.62	5.34	7.27
5	5	3	4.54	5.71	7.54
5	5	4	4.53	5.64	7.77
5	5	5	4.56	5.78	7.98
Large samples			4.61	5.99	9.21

Adapted from *Sturdy Statistics*, by F. Mosteller and R. E. K. Rouke. Copyright 1973 by Addison-Wesley, Reading, Mass. Reprinted by permission.

Answers
to Exercises

Chapter 1
1. (c)   **2.** (b)   **3.** (b)

Chapter 2
1. (b)   **2.** (c)

3.

SCORE	f
5	1
4	1
3	2
2	3
1	4
0	1

4.

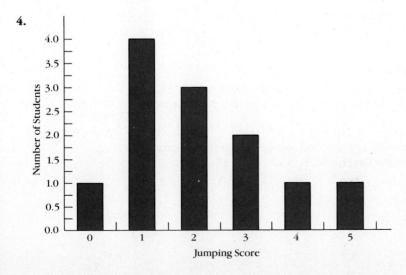

5.

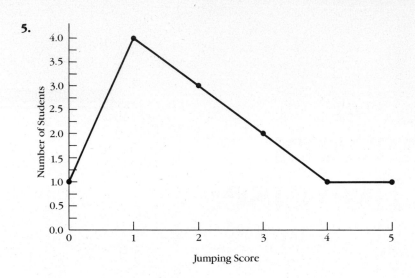

6.

SCORE	CUMULATIVE f
5	12
4	11
3	10
2	8
1	5
0	1

7. 239.5; 219.5

Chapter 3
1. 70.01   **2.** 70.05
3. No. All scores occur with the same frequency.
4. 75.0 − 64.1 = 10.9   **5.** 9.51   **6.** 3.08

Chapter 4
1. Fallbrook, .55; Carlsbad, −.26; San Dieguito, 1.62   **2.** 43
3. The upper quartile represents the top 3.5 schools because: $Pr = B/N \times 100$
$$.75 = X/14$$
$$X = 10.5.$$

So San Dieguito, Poway, and Ramona are in the upper quartile.
4. Just Mountain Empire   **5.** 67.75 to 72.65

Chapter 5
1. (b)   **2.** (a)

3.

4. Unique sequences: $2^7 = 128$.

5. $_5P_3 = \dfrac{5!}{(5-3)!} = \dfrac{5!}{2!} = \dfrac{5 \times 4 \times 3 \times \cancel{2} \times \cancel{1}}{\cancel{2} \times \cancel{1}} = 60$;

60 permutation $\times$ 5 animals = 300 animals

6. $\dfrac{1}{13,123,110} = 7.6 \times 10^{-8}$; $_{10}C_7$; $_{28}C_{10}$

$_{10}C_7 = \dfrac{10!}{7!(10 \times 7)!} = \dfrac{10 \times \cancel{9}^3 \times \cancel{8}^4}{\cancel{3} \times \cancel{2} \times 1}$;

$\dfrac{10 \times 3 \times 4}{14 \times 3 \times 13 \times 5 \times 23 \times 11 \times 19} = \dfrac{4}{7 \times 13 \times 23 \times 11 \times 19} =$

$\dfrac{4}{437,437} = 9.1 \times 10^{-6}$

Alternate approach:

$\dfrac{10}{28} \times \dfrac{9}{27} \times \dfrac{8}{26} \times \dfrac{7}{25} \times \dfrac{6}{24} \times \dfrac{5}{23} \times \dfrac{4}{22} \times \dfrac{3}{21} \times \dfrac{2}{20} \times \dfrac{1}{19} =$

$\dfrac{1}{13 \times 5 \times 3 \times 23 \times 11 \times 7 \times 2 \times 19} = \dfrac{1}{13,123,110}$

Chapter 6

2. No; it must also be representative.

3. A parameter is a numerical characteristic of a whole population; a statistic is a numerical characteristic of a sample from the population.

4. 0.5   **5.** 3.8   **6.** 54.1 to 70.7   **7.** 15.43 to 17.37

8. 13 degrees of freedom; 8.68 to 15.72

Chapter 7

1. No. The mean for the Palo Alto children was 116.9. The probability is less than 1/1000 (or .001) that the observed mean could have differed from the expected value (100) by chance alone.

2. *Null hypothesis:* H_0: $\mu_1 = \mu_2$ (dogs will eat the same amount of each food); *Alternative hypothesis:* H_1: $\mu_1 \neq \mu_2$ (dogs will eat significantly more of one dog food than the other), $df = 18$, $t = \dfrac{5.60}{2.09} = 2.67$;

Decision: Reject null hypothesis: dogs prefer *A*.

3. H_0: $\mu_1 < \mu_2$

Null: The incentive program does not increase weight loss, and greater weight loss by the incentive group is due to chance.

H_1: $\mu_1 > \mu_2$ (acceptable $\mu_1 < \mu_2$ if loss is coded as negative)

Alternative: The incentive program produces significantly more weight loss than control condition without special incentives.

Decision rule: If $t > 1.761$ ($p < .05$, one-tailed), the alternative hypothesis will be accepted; otherwise retain null.

4. H_0: $\mu_1 > \mu_2$
Null: Running in sweatpants does not slow performance.
H_1: $\mu_1 < \mu_2$
Alternative: Running in sweatpants significantly slows performance.
Decision rule: If $t \leq 1.83$, retain null; if $t > 1.83$ ($p < .05$, one-tailed), the alternative hypothesis will be accepted.
$df = 9$; $t = 1.66$, retain null hypothesis: Athletes did not run significantly slower with sweatpants.

5. .05 (5%)

6. A type I error occurs when a researcher concludes that a true difference exists between conditions when in fact the difference was due to chance (random variability). A type II error occurs when a researcher concludes that there is no difference between conditions when in fact there really is a difference. The probability of a type I error is decided by the researcher's alpha level; by setting the alpha level (and probability of a type I error) higher or lower, the researcher affects the probability of a type II error. The power of a test is the probability of finding a true difference between conditions; the power of a test equals one minus β (the probability of a type II error). The power of a test is indirectly related to the probability of a type I error: the lower the probability of a type I error, the higher the chances of a type II error.

7. Yes, it would be important to have a high power for the test because it would be unlikely that a second study would be conducted if a first study erroneously showed that the procedure was ineffective.

Chapter 8

1. $\dfrac{(5)(5-1)}{2} = \dfrac{20}{2} = 10$ t-tests required.

At a .05 alpha level, there is a 50% chance that one of the 10 t-tests will show a significant difference because of chance factors (a type I error).

2.

	SCHOOL #1			SCHOOL #2			SCHOOL #3	
S	$X_{.1}$	$X_{.1}^2$	S	$X_{.2}$	$X_{.2}^2$	S	X_3	X_3^2
X_{11}	2	4	X_{12}	1	1	X_{13}	6	36
X_{21}	3	9	X_{22}	3	9	X_{23}	5	25
X_{31}	1	1	X_{32}	4	16	X_{33}	4	16
X_{41}	2	4	X_{42}	2	4	X_{43}	2	4
X_{51}	3	9	X_{52}	1	1	X_{53}	6	36
X_{61}	4	16	X_{62}	5	25	X_{63}	4	16
X_{71}	1	1	X_{72}	5	25	X_{73}	7	49
X_{81}	1	1	X_{82}	4	16	X_{83}	1	1
X_{91}	5	25	X_{92}	5	25	X_{93}	6	36
X_{101}	1	1	X_{102}	6	36	X_{103}	7	49
$\Sigma X_{.1}$	23		$\Sigma X_{.2}$	36		$\Sigma X_{.3}$	48	
$\Sigma X_{.1}^2$		71	$\Sigma X_{.2}^2$		158	$\Sigma X_{.3}^2$		268
$N_1 = 10$			$N_2 = 10$			$N_3 = 10$		

STEP
1: $\Sigma X_{..} = \Sigma X_{.1} + \Sigma X_{.2} + \Sigma X_{.3} = 23 + 36 + 48 = 107$
2: $\Sigma X_{ij}^2 = \Sigma X_1^2 + \Sigma X_2^2 + \Sigma X_3^2 = 71 + 158 + 268 = 497$
3: $\Sigma X_{..}/N = 107 \div 30 = 3.57$
4: $(\Sigma X_{.1})^2/N_1 = (23)^2 \div 10 = 52.9$
5: $(\Sigma X_{.2})^2/N_2 = 1296 \div 10 = 129.6$
6: $(\Sigma X_{.3})^2/N_3 = 2304 \div 10 = 230.4$
7: $\Sigma(\Sigma X_{.j})^2 \div N_j = 52.9 + 129.6 + 230.4 = 412.9$
8: $(\Sigma X_{..})^2/N = (107)^2 \div 30 = 11{,}449 \div 30 = 381.63$
9: $SS_B = 412.9 - 381.6 = 31.3$
10: $SS_W = 497 - 412.9 = 84.1$
11: $SST = 497 - 381.6 = 115.4$
12. $J - 1 = 3 - 1 = 2$
13: $N - J = 30 - 3 = 27$
14: $N - 1 = 30 - 1 = 29$
15: $MS_B = 31.3 \div 2 = 15.7$
16: $MS_W = 84.1 \div 27 = 3.1$

17: $F(2/27) = \dfrac{15.7}{3.1} = 5.1$
$p < .025$

The null hypothesis: Following the smoking prevention interventions, differences in attitudes toward smoking between the three schools that had the three different programs were nonsignificant. (Alternatively: The smoking intervention programs had no significant effect on attitudes toward smoking.)

3. Yes; the Fisher LSD test *is* appropriate.

$$t = \frac{\bar{X}_1 - \bar{X}_2}{\sqrt{MS_W(1/N_1 + 1/N_2)}} = \frac{2.3 - 3.6}{\sqrt{3.1(2/10)}} = \frac{-1.3}{\sqrt{0.62}} = \frac{-1.3}{0.797} = 2.44, p < .05, 27\ df$$

$$t = \frac{X_1 - X_3}{0.787} = \frac{2.3 - 4.8}{0.787} = \frac{-2.5}{0.787} = 3.18, p < .01, 27\ df$$

$$t = \frac{X_2 - X_3}{0.787} = \frac{3.6 - 4.8}{0.787} = \frac{-1.2}{0.787} = 1.52, p > .05, 27\ df$$

The results of the *t*-test show no significant difference between the groups that received pamphlets about the evils of smoking and those with no intervention. The group with a social learning intervention had significantly more negative attitudes about smoking than either of the other two groups.

Chapter 9

1.

SOURCE	SS	df	MS	F
A	26.28	1	26.28	6.57**
B	0.78	1	0.78	0.20
A × B	270.28	1	270.28	67.57**
Within/error	111.88	28	4.00	
Total	409.22	31		

$**p < .01$

STEP

1: $\Sigma X_{.11} = 68; \Sigma X_{.21} = 36; \Sigma X_{.12} = 24; \Sigma X_{.22} = 85$

2: $N_{.11} = 8; N_{.12} = 8; N_{.21} = 8; N_{.22} = 8$

3: $\Sigma X_{.1.} = \Sigma X_{.11} + \Sigma X_{.12} = 68 + 24 = 92$
$\Sigma X_{.2.} = \Sigma X_{.21} + \Sigma X_{.22} = 36 + 85 = 121$

4: $\Sigma X_{..1} = \Sigma X_{.11} + \Sigma X_{.21} = 68 + 36 = 104$
$\Sigma X_{..2} = \Sigma X_{.12} + \Sigma X_{.22} = 24 + 85 = 109$

5: $N_{a1} = 16; N_{a2} = 16; N_{b1} = 16; N_{b2} = 16$

6: $\Sigma X_{...} = 92 + 121 = 213; 104 + 109 = 213$

7: $N = 16 + 16 = 32$

8: $(\Sigma X_{..})^2/N = (213)^2/32 = 45,369 \div 32 = 1,417.78$

9: $6^2 + 7^2 + 8^2 + 8^2 + 10^2 + 8^2 + 11^2 + 9^2 + 7^2 + 6^2 + 3^2 +$
$1^2 + 6^2 + 7^2 + 4^2 + 2^2 + 3^2 + 2^2 + 4^2 + 3^2 + 6^2 + 4^2 + 1^2 +$
$1^2 + 12^2 + 10^2 + 9^2 + 8^2 + 15^2 + 10^2 + 9^2 + 12^2 = 36 + 49 +$
$64 + 81 + 100 + 64 + 121 + 81 + 49 + 36 + 9 + 1 + 36 +$
$49 + 16 + 4 + 9 + 4 + 16 + 9 + 36 + 16 + 1 + 1 + 144 +$
$100 + 81 + 64 + 225 + 100 + 81 + 144 = 1827$

10: $SS_{\text{Total}} = 1827 - 1417.78 = 409.22$

11(a): $\dfrac{(92)^2}{16} + \dfrac{(121)^2}{16} = 529 + 915.06 = 1444.06$

(b): $SS_A = 1444.06 - 1417.78 = 26.28$

12(a): $\dfrac{(104)^2}{16} + \dfrac{(109)^2}{16} = 676 + 742.56 = 1418.56$

(b): $SS_B = 1418.56 - 1417.78 = 0.78$

13(a): $68^2 + 36^2 + 24^2 + 85^2 = 4624 + 1296 + 576 + 7225 = 13721$

(b): $13721/8 = 1715.12$

(c): $1715.12 - 1417.78 - 0.78 - 26.28 = 270.28 = SS_{AB}$

14: $SS_{\text{Error}} = 409.22 - 26.28 - 0.78 - 270.28 = 111.88$

15: See before step 1.

16: $df_A = 2 - 1 = 1; df_B = 2 - 1 = 1; df_{A \times B} = (1)(1) = 1;$
$df_{\text{within/error}} = 32 - (2 \times 2) = 32 - 4 = 28; df_{\text{Total}} = 32 - 1 = 31$

17: (See table): $111.88 \div 28 = 3.9957 = 4.00$

18: (See table): $F_A = 26.28 \div 4 = 6.57; F_B = 0.20; F_{A \times B} = 67.57$

19: $F_A(1, 28) = 6.57, p < .05; F_B(1, 28) = 0.20, p > .05;$
$F_{A \times B}(1, 28) = 67.57, p < .01$

20:

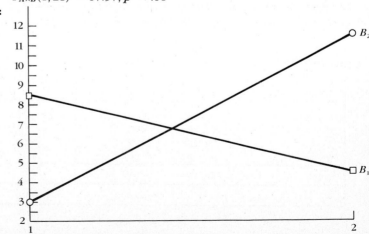

2. STEP **1:** $\Sigma X_{.11} = 7 + 9 + \cdots + 9 = 99; \Sigma_{.21} = 145; \Sigma_{.12} = 16; \Sigma_{.22} = 42$

2: $N_{.11} = 10; N_{.12} = 10; N_{.21} = 10; N_{.22} = 10$

3: $\Sigma X_{.1.} = 115; \Sigma_{.2.} = 187$

4: $\Sigma X_{..1} = 244; \Sigma_{..2} = 58$

5: $N_{a1} = 20; N_{a2} = 20; N_{b1} = 20; N_{b2} = 20$

6: $\Sigma X_{...} = 115 + 187 = 302; 244 + 58 = 302$

7: 40

8: $(\Sigma X_{...})^2/N = (302)^2 \div 40 = 2280.1$

9: $49 + 81 + 324 + 16 + 36 + 81 + 121 + 196 + 144 + 81 +$
$256 + 324 + 441 + 81 + 256 + 289 + 169 + 196 + 144 +$
$81 + 9 + 0 + 1 + 4 + 0 + 4 + 0 + 16 + 1 + 9 + 36 + 16 +$
$25 + 1 + 4 + 49 + 25 + 4 + 16 + 36 = 3622$

10: $SS_{Total} = 3622 - 2280.1 = 1341.9$

11: $\dfrac{115^2}{20} + \dfrac{187^2}{20} = 661.25 + 1748.45 = 2409.70;$
$SS_A = 2409.70 - 2280.1 = 129.6$

12: $244^2/20 + 58^2/20 = 2976.8 + 168.2 = 3145;$
$SS_B = 3145 - 2280.1 = 864.9$

13: $99^2 + 145^2 + 16^2 + 42^2 = 9,801 + 21,025 + 256 + 1764 +$
$32,846 \div 10 = 3,284.6$
$3,284.6 - 864.9 - 129.6 - 2,280.1 = 10.00$

14: $1341.9 - 129.6 - 2378.7 - (-1503.8) = 337.4$

SOURCE	SS	df	MS	F
A	129.6	1	129.6	13.83**
B	864.9	1	864.9	92.28**
A × B	10.00	1	10.00	1.07
Within/Error	337.40	36	9.37	
Total	1341.9	39		

$**p < .01$

16: $1, 1, 1, 40 - (2 \times 2) = 36; 40 - 1 = 39$

17: See above.

18: $F_A = 129.6/9.37 = 13.83$
$F_B = 864.9/9.37 = 92.28$
$F_{A \times B} = 10/9.37 = 1.07$

19: $(1, 36)$
Using 40, $p = .05$ at 4.08; $p = .01$ at 7.31. The F's for A and for B are significant at the .01 level. The interaction is nonsignificant.

20:

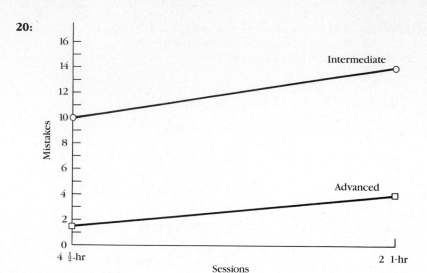

Chapter 10

1. STEP

1: $N = 16$

2: $\Sigma X = 257 + 271 + \cdots + 260 = 4429$

3: $\Sigma Y = 240 + 269 + \cdots + 251 = 4229$

4: $\Sigma X^2 = 66,049 + 73,441 + 48,400 + 106,276 + 48,841 + 76,176$
$+ 95,481 + 48,400 + 78,400 + 86,436 + 112,896 + 85,849$
$+ 66,049 + 50,176 + 148,225 + 67,600$
$= 1,258,695$

5: $\Sigma Y^2 = 57,600 + 72,361 + 50,625 + 56,169 + 48,841 + 75,076$
$+ 108,241 + 40,804 + 67,600 + 63,001 + 108,241 + 82,944$
$+ 68,644 + 40,000 + 152,881 + 63,001$
$= 1,156,029$

6: $\Sigma XY = (240 + 257) + \cdots$, etc. $= 61,680 + 72,899 + 49,500$
$+ 77,262 + 48,841 + 75,624 + 101,661 + 44,440 + 72,800$
$+ 73,794 + 110,544 + 84,384 + 67,334 + 44,800 + 150,535$
$+ 65,260$
$= 1,201,358$

7: $(\Sigma X)^2 = (4429)^2 = 19,616,041$

8: $(\Sigma Y)^2 = (4229)^2 = 17,884,441$

9: $N\Sigma XY = 16(1,201,358) = 19,221,728$

10: $(\Sigma X)(\Sigma Y) = (4429)(4229) = 18,730,241$

11: $N\Sigma XY - (\Sigma X)(\Sigma Y) = 491,487$

12: $N\Sigma X^2 = (16)(1,258,695) = 20,139,120$

13: $N\Sigma X^2 - (\Sigma X)^2 = 523,079$

14: $\dfrac{491,487}{523,079} = 0.9396 = .94$ **15:** $\bar{X} = \dfrac{4429}{16} = 276.81$

16: $\bar{Y} = \dfrac{4229}{.16} = 264.31$

17: $b\bar{X} = 260.20$ **18:** $a = 264.31 - 260.20 = 4.11$

19: $Y = a + bx = 4.11 + .94X$

2. Regression would be used to translate raw mathematics scores into estimated raw reading scores. If you wanted just a numerical expression of the relationship between mathematics and reading scores, correlation would be used.

3. Steps 1–13 same as in problem #1.

  14: $N\Sigma Y^2 = (16)(1,156,029) = 18,496,464$
  15: $N\Sigma Y^2 - (\Sigma Y)^2 = 18,496,464 - 17,884,441 = 612,023$
  16: $\sqrt{[N\Sigma X^2 - (\Sigma X)^2][N\Sigma Y^2 - (\Sigma Y)^2]} = \sqrt{(523,079)(612,023)}$
    $= \sqrt{320,136,370,000} = 565,806$
  17: $r = \dfrac{491,487}{565,806} = .8686 = .87$

4. Positive   5. $(.87)^2 = .76$ or 76%   6. Coefficient of determination   7. We do not know about causation from a correlation. In this example, a third variable may be causing both.

Chapter 11

1. STEP 1: $N = 12$
  2: $\Sigma X = 105 + 99 + 112 + \cdots + 106 = 1,248$
  3: $\Sigma Y = 107 + 106 + 105 + \cdots + 110 = 1,245$
  4: $\Sigma X^2 = 105^2 + 99^2 + 112^2 + \cdots + 106^2 = 11,025 + 9801 + 12,544$
    $+ 13,456 + 12,321 + 7,396 + 10,816 + 10,404 + 7,569 + 14,400$
    $+ 10,000 + 11,236$
    $= 130,968$
  5: $\Sigma Y^2 = 11,449 + 11,236 + 11,025 + 13,225 + 12,996 + 7,569$
    $+ 9,216 + 9,025 + 8,464 + 12,996 + 10,816 + 12,100$
    $= 130,117$
  6: $\Sigma(XY) = (105 \times 107) + (99 \times 106) + \cdots + (106)(110) = 11,235 +$
    $10,494 + 11,760 + 13,340 + 12,654 + 7,482 + 9,984 + 9,690 +$
    $8,004 + 13,680 + 10,400 + 11,660 = 130,383$
  7: $(\Sigma X)^2 = (1248)^2 = 1,557,504$
  8: $(\Sigma Y)^2 = (1245)^2 = 1,550,025$
  9: $N\Sigma XY = (12)(130,383) = 1,564,596$
  10: $(\Sigma X)(\Sigma Y) = (1248)(1245) = 1,553,760$
  11: $N\Sigma XY - (\Sigma X)(\Sigma Y) = 1,564,596 - 1,553,760 = 10,836$
  12: $N\Sigma X^2 = (12)(130,968) = 1,571,616$
  13: $N\Sigma X^2 - (\Sigma X)^2 = 1,571,616 - 1,557,504 = 14,112$
  14: $N\Sigma Y^2 = (12)(130,117) = 1,561,404$
  15: $N\Sigma Y^2 - (\Sigma Y)^2 = 1,561,404 - 1,550,025 = 11,379$
  16: $\sqrt{(\text{Step 13})(\text{Step 15})} = 12,672.0$
  17: $r = \dfrac{\text{Step 11}}{\text{Step 16}} = \dfrac{10,836}{12,672} = 0.8551 = .86$

    Thus the test–retest reliability is .86.

2. Measurement error is the difference between the theoretical true score for quality we wish to measure and the actual value obtained using a particular measurement tool.

3. STEP 1: $NS = 100$
  2: $p = .41, .86, .16, .42, .98, .71, .16, .25, .91, .46$

3: $q = .59, .14, .84, .58, .02, .29, .84, .75, .09, .54$

4: $pq = .24, .12, .13, .24, .02, .21, .13, .19, .08, .25$

5: $\Sigma pq = 1.61$

6: Given as 3.65

7: $3.65 - 1.61 = 2.04$

8: $\dfrac{2.04}{3.65} = .56$

9: 10

10: $\dfrac{N}{N-1} = \dfrac{10}{9} = 1.11$

11: $(.56)(1.11) = .62 = KR_{20}$

4. $N = \dfrac{r_d(1 - r_o)}{r_o(1 - r_d)} = \dfrac{.90(1 - .81)}{.81(1 - .90)} = \dfrac{(.90)(.19)}{(.81)(.10)} = \dfrac{.171}{.081} = 2.11$

$(2.11)(40) = 84.4$ items or 85 total items $- 45$ already on test 45 new ones.

5. $\hat{r}_{12} = \dfrac{r_{12}}{r_{11}r_{12}} = \dfrac{.63}{(.70 \times .59)} = \dfrac{.63}{.413} = \dfrac{.63}{.6427} = .98$

6. $r_{12}\text{max} = r_{11}r_{12} = (.5)(.6) = .3 = .55$

Chapter 12

1.

	EAST	SOUTH	MIDWEST	WEST	TOTAL
1. Observed (O)	43	19	21	17	100
2. Expected (E)	25	25	25	25	100
3. $O - E$	18	-6	-4	-8	0
4. $(O - E)^2$	324	36	16	64	440
5. $\dfrac{(O - E)^2}{E}$	12.96	1.44	0.64	2.56	17.60

$\chi_2 = \Sigma[(O - E)^2/E] = 17.60$, critical value at .05 level, 7.8; $df = 3$. Actually, 17.60 is significant at the .001 level. The regions are not equally represented.

2.

	PASS	FAIL	TOTAL
Men	40	42	82
Women	47	22	69
Total	87	64	151

$87/151 = .58$ pass; $64/151 = .42$ fail; $(.58)(82) = 48$ men expected to pass; $(.58)(69) = 40$ women expected to pass

	MEN PASS	WOMEN PASS	MEN FAIL	WOMEN FAIL	TOTAL
O	40	47	42	22	151
E	48	40	34	29	151
$O - E$	-8	7	8	-7	0
$(O - E)^2$	64	49	64	49	226
$(O - E)^2/E$	1.33	1.22	1.88	1.69	6.12

There is really only one degree of freedom here. The null hypothesis is that men and women have an equal chance of passing the bar exam; the alternative hypothesis is that men pass at a significantly different rate than women. The null hypothesis is rejected at the .025 level of significance.

3.

STEP 1:

A19	1
A16	1
A14	1
B12	1
A/11, B/11	2
B10	1
B8	1
A7	1
A6	3
A4	1
A3	1
B2	1
B0	1

$$\frac{N + 1}{2} = \frac{16 + 1}{2} = \frac{17}{2} = 8.5$$

2: 7.5 = median

3:

	AT OR ABOVE	BELOW	
Team A	A 4	B 4	A + B = 8
Team B	C 4	D 4	C + D = 8

4: $\chi^2 = \dfrac{N(AD - BC)^2}{(A + B)(C + D)(A + C)(B + D)} = \dfrac{16(16 - 16)2}{(8)(8)(8)(8)} = 0$

5: $df = 1$

6: According to median test, both from a single population have same median.

4. *Null:* There is no difference in scoring rank between teams A and B. [They belong to a common distribution.] *Alternative:* There is a significant difference in scoring ranks between the two teams that is unlikely to have occurred by chance.

SCORE	19	16	14	12	11	11	10	8	7	6	6	6	4	3	2	0
RANK	1	2	3	4*	5.5*	5.5	7*	8*	9	11	11*	11*	13	14	15*	16*

STEP 3: $R_1 = 1 + 2 + 3 + 5.5 + 9 + 11 + 13 + 14 = 58.5$

4: $R_2 = 4 + 5.5 + 7 + 8 + 11 + 11 + 15 + 16 = 77.5$

5: $N_1 N_2 = (8)(8) = 64$

6: $N_1(N_1 + 1) = 8(9) = 72$

7: $72 \div 2 = 36$

8: $N_1 N_2 + \dfrac{N_1(N_1 + 1)}{2} - R_1 = 64 + 36 - 58.5 = 100 - 58.5 = 41.5$

9: $U = \dfrac{N_1 N_2}{2} = \dfrac{64}{2} = 32$

10: $N_1 + N_2 + 1 = 8 + 8 + 1 = 17$

11: $U = \dfrac{N_1 N_2 (N_1 + N_2 + 1)}{12} = \dfrac{64(12)}{12} = 90.67 = 9.522$

12: $Z = \dfrac{U_1 - U_E}{u} = \dfrac{41.5 - 32}{9.522} = \dfrac{9.5}{9.522} = 0.998$

13: $0.998 < 1.96$. The observed difference is not statistically significant; null hypothesis retained.

5. STEP 1:

DRUG 1		DRUG 2		PLACEBO	
0	20.5	1	18	6	7.5
2	15	3	12.5	21	1
4	10	4	10	17	2
1	18	6	7.5	9	4
0	20.5	4	10	2	15
3	12.5	2	15	1	18
7	6	8	5	16	3
$R_1 = 102.5$		$R_2 = 78$		$R_3 = 50.5$	

2:

SCORE 21 17 16 9 8 7 6 6 4 4 4 3 3 2 2 2 1 1
1 0 0

RANK 1 2 3 4 5 6 7.5 7.5 10 10 10 12.5 12.5 15 15
15 18 18 18 20.5 20.5

3: $R_1^2 = (102.5)^2 = 10506.25$
$R_2^2 = (78)^2 = 6084$
$R_3^2 = (50.5)^2 = 2550.25$

4. $R_1^2/n_1 = 1500.89$
$R_2^2/n_2 = 869.14$
$R_3^2/n_3 = 364.32$

5: $\Sigma\,(R_i^2/n_i) = 2734.35$

6: $12/N(N + 1) = 12/21(21 + 1) = 12/462 = 0.025974$

7: $3(N + 1) = 3(21 + 1) = 66$

8: $H = \dfrac{12}{N(N + 1)} \sum \dfrac{R_i^2}{n_i} - 3(N + 1) = [(0.025974)(2734.35)] - 66$
$= 71.02 - 66 = 5.02$

9: $df = 3 - 1 = 2$

10: 6.0 is required ($5.02 < 6.0$) so the difference is not significant at the .05 level.

The null hypothesis, that the groups came from the same distribution, is retained.

6. A test has high relative efficiency when it needs relatively few subjects to attain a certain power.

Index

A

α (alpha), definition of, 161
a, definition of, 219
Abscissa (x-axis), 28
  definition of, 37
Absolute 0, 17
  definition of, 37
Additive model, 195–197
  definition of, 195, 209
  graph representation of,
    196–197
Additive rules, application of,
  90–91
Alienation, coefficient of, 237, 240
  definition of, 241
Alpha (α), definition of, 161
Alternative hypothesis, 111–112
  for chi-square:
    with frequency data, 268,
      269, 271
    with 2 × 2 contingency
      table, 275, 276, 279, 280
  in correlation analysis, 227
  general statement of, 143
  for Mann–Whitney U-test, 286
  for median test, 282, 284
  for one-way ANOVA, 173, 176,
    183, 184
  sampling distribution of sample
    means under, 162–163
  symbol for, 144
Analysis of variance (ANOVA):
  alternative hypothesis for, 173,
    176, 183, 184

assumptions of, 295
null hypothesis for, 173, 176,
  183, 184
one-way:
  assumptions of, 177
  and between-groups sums of
    squares, 178–179, 180,
    197
  computational formulas for,
    183, 184
  and degrees of freedom, 176,
    179–180, 181, 182–183
  example data for, 172–173
  F for, 181, 182, 198
  grand mean in, 174
  indexing subscripts in,
    177–178
  mean squares in, 180–182,
    197–198
  multiple-comparisons prob-
    lem in, 183, 187–188
  negative scores in, 183
  notational system for,
    177–178
  partitioning in, 174–176, 197
  and post hoc tests, 187–188
  as ratio of between- to
    within-groups variance,
    173–174, 175–176, 188
  robustness of, 177
  sample sizes for, 176, 177
  steps in calculating, 182–186
  table summary of, 180–183
  and total sums of squares,
    178, 180, 197

total variance in, 174, 175
and unequal N, 177
usefulness of, 169–170, 172,
  191
and within-groups sums of
  squares, 178–179, 180,
  197
two-way:
  and additive model, 195
  assumptions of, 208–209
  and between-groups sums of
    squares, 199–201,
    202–204
  contingent effects in, 195, 197
  and degrees of freedom, 205
  example data for, 182,
    193–194, 200–201, 206
  F for, 198, 205–206, 207–208
  graph representation of,
    195–196, 206, 209
  interactions in, 195–197
  main effects in, 194–197
  mean squares in, 205
  partitioning in, 198–201
  questions evaluated by, 194,
    209
  as ratio of between- to
    within-groups variance,
    197–198
  robustness of, 208
  and sampling error, 197–198
  steps in calculating, 198–208
  summary table for, 204
  and total sums of squares,
    199–201, 202–204

343

One-Way ANOVA

Between groups SS	$\sum \left[\dfrac{(\sum X_{\cdot j})^2}{n_j} \right] - \dfrac{(\sum X_{\cdot\cdot})^2}{N}$	$df = J - 1$ $MS = SS_B/df_B$	$F = MS_B/MS_w$ p: See Appendix 3.
Within groups SS	$\sum \left[(\sum X_{\cdot j}^2) \right] - \dfrac{(\sum X_{\cdot j})^2}{n_j}$ or $\text{Total}_{SS} - \text{Between}_{SS}$	$df = N - J$ $MS = SS_w/df_w$	
Total SS	$\sum X_{ij}^2 - \dfrac{(\sum X_{ij})^2}{N}$	$df = N - 1$	